GEEK PHYSICS

"This book won my heart after it estimated lightsaber temperature from the color of molten metal."

—ZACH WEINERSMITH
author and illustrator of the webcomic *Saturday Morning Breakfast Cereal*

"*Geek Physics* shows us the joy to be found in using simple models and physics principles to dig deeper into everything from sports to comic-book movies. And, as it turns out, adding a little physics makes everything more fun."

—CHAD ORZEL
author of *How to Teach Quantum Physics to Your Dog*

"Everything that happens in the world is described by physics. Interestingly, even things that don't happen are described by physics. In this delightful book, Rhett Allain uncovers the science behind some of the most fun hypothetical questions we can ask, from Han shooting first to the power in Superman's fists."

—SEAN CARROLL
theoretical physicist and author of *The Particle at the End of the Universe*

"*Geek Physics* will cause you to see the relevance of physics to life's hidden, everyday questions. It's the superposition of Hollywood, Mythbusters, YouTube, physics and Rhett Allain's knack for asking interesting questions."

—DR. AARON TITUS
codeveloper of WebAssign

GEEK PHYSICS

ALSO BY RHETT ALLAIN

Just Enough Physics

The Key Ideas in Physics to Get You Through Your Introductory Class

National Geographic
Angry Birds™ Furious Forces!

The Physics at Play in the World's Most Popular Game

GEEK PHYSICS

Surprising Answers to the Planet's Most Interesting Questions

Rhett Allain

WILEY

Wiley General Trade, an imprint of Turner Publishing Company

424 Church Street • Suite 2240 • Nashville, Tennessee 37219

445 Park Avenue • 9th Floor • New York, New York 10022

www.turnerpublishing.com

Geek Physics: Surprising Answers to the Planet's Most Interesting Questions

Cover design: Susan Olinsky

Book design: Taylor Reiman

Front cover image: ©iStock.com/BunnyHollywood

Interior illustrations: ©Rhett Allain

Library of Congress Cataloging-in-Publication Data

Allain, Rhett, author.
  Geek physics : surprising answers to the world's most interesting questions / Rhett Allain.
    pages cm
  ISBN 978-1-68162-049-7

1. Physics--Miscellanea. 2. Physics--Problems, exercises, etc. 3. Physics--Humor. I. Title.
QC75.A45 2015
  530--dc23

                              2015000285
Printed in the United States of America
10 9 8 7 6 5 4 3 2 1

CONTENTS

ACKNOWLEDGMENTS

First, a big thank you to my wife, Ashley,
for putting up with my excessive blogging and instantaneous
physics calculations.
I would also like to acknowledge my kids with their
mostly unknown contributions to both my blog and book.

If I ever needed a stand-in hand model, they were there.
Finally, I would like to thank Eric Nelson for coming up with the idea
to put these blog posts into a book and getting me into this mess.

INTRODUCTION

Scholars may forever debate the distinction between "geek" and "nerd." For me, there is no distinction. They are proud titles of a proud group of men and women dedicated to a particular field. Maybe one geek is an expert in all the different types of insects that can devour a dead cow. Perhaps a nerd can identify all the different lightsabers that the Jedi use. Geeks can build cool things or write interesting narratives. Some are a lot like me: they take things apart to figure out how they work but can't always put them back together.

I think the geek and nerd labels are no longer associated with negative things. People like to be called geeks. Geek is cool. Just look at popular television shows like *The Big Bang Theory* or *MythBusters*. We all have at least a small inner-geek and inner-nerd in us, and that's okay.

What about me? Am I a geek? Sure. When I was younger, I was very interested in geeky things: space, comic books, science fiction, and building things. One of the early projects I worked on in my teenage years was a small electric device that used two photocells and a motor to track the motion of a light. I loved that little thing. I later went on to study physics and eventually became a faculty member at Southeastern Louisiana University. Of course, I still like those same geeky things from my past. What could be better than combining physics and geek culture? Maybe a good peanut butter and jelly sandwich would be better, just maybe.

So that's what I do. I use two basic ideas to look at the world around me. First, there are the fundamental ideas of physics such as the momentum principle, the work-energy principle, or the relationship between electric and magnetic fields (we call these relationships Maxwell's Equations). But that's not all I use in my analysis of both real and fictional phenomena. The second is model building. What is a model? Is it like the car model I built as a kid? Yes, it's exactly the same. A model can be anything used to represent something else. The gravitational model creates a mathematical expression relating the mass of two objects to the force between them. We can also build models which aren't already established. How fast does a tweet move across the country? Just collect some data and find an expression that works and you're done. By answering those questions, you've just built a model. The model might be complicated, but the idea of model building is straightforward.

Why would anyone waste their time looking at the physics of *Star Wars*? Why would anyone care how many fish Gollum (from *The Hobbit*) needs to eat to maintain his body temperature? If I calculate the perfect amount of ice to cool off my beer, is that actually how much I will use? Don't I have more important things to do? Well, what is "wasting time"? Is watching a movie wasting time? What about reading a book or painting a picture? What one might consider wasting time, another might consider it being productive.

Is all this analysis of unrealistic things useless? How about an analogy? Consider the popular video game *Angry Birds™* as an example. Personally, I love this game but I find the most enjoyable aspect of it is to try and figure out how it works. Do the birds move like they would on the Earth? Are there different laws of physics in the *Angry Birds™* game? The manner in which I can analyze the way the game operates is a lot like physics research in the real world. I can use the same methods and similar tools on video games that I would use in other areas of physics. Using physics to analyze a video game is a lot like climbing up a modular rock climbing wall. Climbing these walls helps you build your rock climbing skills, but there's nothing to find at the top. If you went to an actual mountain, you would use the exact same rock climbing skills where the only difference is having an actual top to reach. So, to complete the analogy, real physics research is like real mountain climbing. Analyzing the physics of *Angry Birds™* is like climbing a rock wall.

What's next? Well, now we get started. I will walk you through some of my favorite ideas. All of these ideas could be considered a combination of physics and geek culture, thus the name *Geek Physics*.

Before we start, let me add one note:

While answering these awesome questions, I like to share the process of thinking about building these models. This means I will also describe some of the fundamental physics concepts the models are built with. However, some of these fundamental ideas will pop up in more than one question. I have chosen to give a quick introduction to the concept each time it is used. That means multiple chapters will contain the same explanation. Don't worry, this is good for two reasons. First, physics can be complicated and going over something more than once won't hurt you. Second, because the explanations are in more than one location, you can skip around in the book if you like.

CHAPTER 1: ACTUAL PHYSICS

When I was a student in college, we would often be introduced to real physics content by first reading material in an actual textbook. After that, the professor would give a lecture on the topic and maybe work out some examples. Finally, we would go to a lab class where we could play with some equipment and further explore these physics concepts. This approach is fine, but what if there's a better way? What if we start with the experiment first? Let's do that.

I find that this experiment is the most fun if you use your smartphone with the Google search app. Just ask your phone, "Why do astronauts float in space?" When I do this, and it might still work for you, I get this as the official answer:

> **Astronauts float** around in **space** because there is no gravity in **space**. Everyone knows that the farther you get from Earth, the less the gravitational force is. Well, **astronauts** are so far from the Earth that gravity is so small. This is why NASA calls it microgravity.
>
> **Why Do Astronauts Float** Around in **Space**? | WIRED
> www.wired.com/2011/07/**why-do-astronauts-float**-around-in-**space**/   Wired ˅

If you follow that link, it will lead you to my blog. Yes, I wrote that explanation and it is totally wrong. Before you get too upset, you should know that it was wrong on purpose. This Google-approved answer is a common answer many people use to explain weightlessness that I wrote as an introduction to the idea of "weightlessness." So, the Google answer is wrong, but they missed my other common answer: "astronauts are weightless because there is no air in space."

What is wrong with these two very common ideas about gravity? Let's start with the lack of atmosphere.

The Moon is a perfect example of gravity without air. Look at any of the videos from the Apollo Moon landings. If you want a suggestion, search for John Young's famous "jump salute." There is no air on the Moon but the astronauts don't float away. The Moon's gravity still holds them down, yet the Moon's smaller mass means less gravity, which leads us to the other explanation.

Maybe astronauts in space float around because they are too far away from the Earth for the gravitational force to be significant. To answer this, let's look at the gravitational force. The typical model for gravity, as developed by Newton, says that the gravitational force is an attractive interaction between any two objects with mass. The force is proportional to the product of the masses and inversely proportional to the square of the distance between these objects.

As an equation, it would look like this:

$$s_w(t) = (0.0314 \text{ m/year})t + 4.656 \text{ m}$$

The constant G is the universal gravitational constant. It has a value of 6.67 x 10^{-11} Nm²/kg². And what about the famous value of g equal to 9.8 N/kg (often listed at 9.8 m/s²)? That is the value of the gravitational force per mass, just only on the surface of the Earth, it isn't "universal" gravity.

Now, check this out: if I have something sitting on the ground, it is interacting with the Earth. The mass of the Earth is 5.97 x 10^{24} kg and the center of the Earth is 6.38 x 10^6 m (the radius of the Earth) away from the surface, or the ground. Let me put these values into the gravitational model. You can do it yourself as a check.

Do you know what you get? You get a gravitational force of 9.8 newtons per kilograms of mass. Doesn't this expression say that the gravitational force gets weaker as you get farther from the Earth? Yes, but not by as much as you think. A typical height for an orbiting Space Shuttle is about 360 km above the surface of the Earth. Suppose I have a 75 kg astronaut. What would be the weight (gravitational force) on the astronaut both on the surface and in orbit? The only difference will be the distance between the astronaut and the center of the Earth.

If you compare the gravitational force on the astronaut on the surface of the Earth and in orbit, you get 734 newtons (165 pounds) to 657 newtons (147 pounds). Is the gravitational force smaller in orbit? Yes. Enough to call it "weightless"? No. The gravitational force in orbit is 89 percent as large as on the surface. So, this isn't the correct explanation for "weightlessness."

Let's look at how you feel weight in the first place. Let me go ahead and say that what you are feeling right now (assuming you are on the surface of the Earth) isn't really gravity. Here are some scenarios inside an elevator that will help show why:

EXAMPLE 1:   Go stand in an elevator. Do not push the buttons.
Just stand there so that the elevator is at rest.
How do you feel? Awkward? Here is a diagram:

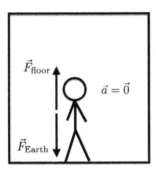

Because you are at rest and staying at rest, you are in equilibrium (acceleration is zero). If your acceleration is zero, the net force must also be zero (technically, the zero vector). The two forces on you are the force from the floor pushing up and the gravitational interaction with the Earth pulling down. The magnitudes of these two forces have to be equal in order for the net force to be zero.

EXAMPLE 2:   Now, push the "up" button. During the short interval
that the elevator accelerates upwards, how do you feel?
Anxious? Or, maybe you feel a tad bit heavier. If your
elevator is like the one in my building, you might feel
frustrated at how slow the damn thing goes.
And what's that funny smell? Here is a diagram for
the upward accelerating elevator (and you):

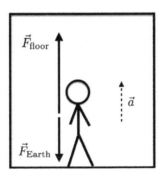

In terms of forces, what has to be different? If the person is accelerating upwards, the net force must also be upwards. Using the same two forces shown above, there are two ways this can happen. The floor can push *more* on you, or the Earth can pull *less*. Because the gravitational force depends on your mass, the Earth's mass, and the distance between those, the gravitational force doesn't change. This means the floor must push harder on you. Interesting that you feel heavier and yet the gravitational force is the same.

EXAMPLE 3:   You are now nearing the top floor and the elevator has to stop. Because it was moving up but slowing down, it must also accelerate in the downward direction.

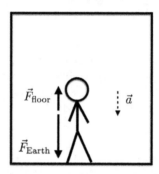

Now the net force must be in the downward direction. Again, the magnitude of the gravitational force doesn't change. The only thing that can happen is for the floor to push less, making you feel lighter.

EXAMPLE 4:   Suppose the elevator cable breaks and the elevator falls. In this case, the acceleration of the elevator will be -9.8 m/s^2 (just like any free-falling object). How much would the floor have to push up on the person to accelerate down at -9.8 m/s^2? It wouldn't have to push at all. The force the floor exerts on you would be zero. How would you feel? You would feel scared. After all, you are in an elevator with the cable cut. How else do you think you would feel? Well, maybe you could be scared *and* hungry if you were late for lunch or something. Oh, and you would feel weightless. Could this really happen? Absolutely! In fact, some people even pay to do this. Look at any number of rides at an amusement park where you're intentionally getting in a car that "drops," like the Tower of Terror.

Let me summarize what we have so far:

- In all these situations,
  the gravitational force does not change.

- For the different situations,
  you have different accelerations.

- The less the floor pushes on you,
  the lighter you feel.

- If the floor doesn't push on you at all,
  you feel weightless.

Moving away from the elevator scenarios, let's look at one final example of weightlessness on the Earth: the vomit comet. Yes, it's real. The vomit comet is a plane that flies in a manner that gives it a downward acceleration just like the falling elevator. However, unlike the elevator, the vomit comet doesn't crash into the ground. In order to prevent a crash, the plane alternates between weightlessness and gaining altitude. These maneuvers have a tendency to cause motion sickness in quite a few passengers, giving the plane the nick-name of "vomit comet."

In the movie *Apollo 13*, the weightless scenes were filmed inside the vomit comet. This way, the scenes not only looked weightless but *were* weightless. Of course, this meant scenes were shot in chunks of about thirty seconds at a time.

Now, returning to space, we can see that the astronauts are in the Space Shuttle which is in orbit around the Earth. But is it accelerating? Yes, it is accelerating because the Earth pulls on it through the gravitational force. Even though it is moving in a circle, it is still accelerating. You could say that the Space Shuttle is indeed falling because its motion is determined by the gravitational force. However, because it doesn't necessarily get closer to the Earth during its motion, it makes more sense to call it "in orbit."

Think about it this way: have you ever swung a bucket of water in a vertical circle hard enough that the water doesn't come out? Rather than spilling out, the circular acceleration causes the water to press against the bottom of the bucket. Now, imagine if your personal gravity was strong enough to pull the water away from the bottom of the bucket—this is what is happening to the astronaut. Astronauts are just like the water in the bucket. Instead of the bottom of the bucket pushing them towards the Earth, it is the gravitational force pulling them.

What if you are actually in a place where the gravitational force is zero (like a place far away from other massive objects)? Can you make it feel like you have weight in this case, as most sci-fi movies show? Yes, you can. Think back to Example 2 where the elevator was accelerating up. If you were in a place with no gravity but were in a moving elevator, you would feel the acceleration. You wouldn't feel weightless anymore. This is essentially the opposite of what is occurring in the orbital case. If you can make the spacecraft accelerate with a magnitude of 9.8 m/s^2, it will feel just like you are on the Earth.

Constantly increasing the speed of a spaceship with rockets will give the feeling of a gravitational force. But maybe you want to just stay in orbit around the Earth rather than travel to another planet. Is there some other way you could have an acceleration to produce an apparent weight? Well, you could make a spaceship that spins. By moving in a circle (on the inside of the spaceship), you would have an acceleration and thus a net force. If you spun water in a bucket in zero gravity, it would still stay in the bucket. If you were in the bucket, it would feel like gravity. The floor pushes up on the astronaut with the same magnitude as standing in a stationary elevator here on the Earth. These two situations would essentially feel the same (but not exactly the same because the top of the rotating head of the astronaut is moving differently than the feet).

Actually, if you recall the movie *2001: A Space Odyssey* there was a scene showing two people walking around in just such a spinning spaceship—that is the scenario I am proposing here.

How about a quick review. Is there gravity in space? Yes, there usually is unless you are far from any large, planet-sized objects. Astronauts appear weightless because they and their spaceship are both accelerating due to the gravitational force. You can produce the same effect as gravity by making the spaceship accelerate with some external force.

IS IT BETTER TO CRASH INTO ANOTHER CAR, OR A BRICK WALL?

In an episode of *MythBusters*, there was an attempt to simulate two large trucks crashing head-on while smashing a smaller car in between them. Experimentally, this is a difficult thing to set up (especially if you only have two trucks to destroy). For their first test, Adam and Jamie, the two Mythbusters, towed two eighteen-wheeler trucks in such a way that both trucks would be traveling at a speed of 50 mph and crash into a stationary car at the same time. The results were impressive, but the stationary car wasn't completely contained in the collision. As the two trucks collided with the car, it moved out of the space between them to prevent complete destruction.

For the next test, they changed it so that a rocket sled crashed into a stationary car that was next to a stationary wall. One of the MythBusters made the claim that if you crash a car going 100 mph into a wall, it would be the equivalent of two 50 mph cars colliding. Is that true? Intuitively, it kind of makes sense. But wouldn't you rather crash into a wall at 50 mph than hit another car going 50 mph in the opposite direction? That statement garnered so much online protest that *MythBusters* actually addressed it in a future episode.

To reach an answer, let's look at the physics from a different *MythBusters* episode.

The myth was that if you put two phone books together with their pages alternating, then it would be impossible to pull them apart. To test this, Adam and Jamie put the two phone books together as directed and then tried to pull them apart in a sort of tug of war.

Here's a diagram:

It looks like a great experiment, but what's wrong with the way they are pulling? By pulling this way, the MythBusters produced only 320 pounds of force on the book. They could have done twice that.

Let's look at the forces on the phone book for the case in which Adam and Jamie pulled in opposite directions. The book is not changing in speed, so the total force on the book must be zero. Essentially, Jamie is exerting 320 pounds of force on the phone book and Adam is just holding it in place (or vice versa). But what would happen if a cable tied to a wall replaced Adam opposite to Jamie? Nothing would change. A wall can exert a force on the rope just like Adam can. Can walls actually push? Sure they can. Have you ever tried to push a wall? The wall pushes back.

What if Adam and Jamie tried to pull like this?

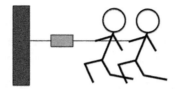

In this case, both Adam and Jamie would pull 320 pounds and the wall would pull the other way with 640 pounds of force. In the show, the MythBusters tried pulling the phone books apart with two cars pulling in opposite directions. They recorded a force of 4,800 pounds. The MythBusters eventually pulled the two phone books apart using two military vehicles that registered a pulling force of 8,000 pounds. If they had instead used the two civilian cars pulling on the same side (with the other side tied to a tree), they would have pulled with a force of 9,600 pounds. This should have been enough to get the phone books apart. But maybe the MythBusters are just smart enough to make it so they had to use the military vehicles. That's what I would have done.

This brings us back to crashing into walls versus crashing into cars. Let me start by looking at two cars in space going the same speed and crashing into an object. Why are they in space? In space I can assume the cars are traveling with no extra forces on them and with a constant velocity. Really, it's just easier to model. I'll assume this is far from any massive objects, so gravitational forces can be neglected. Also, there's no air resistance or friction because the cars aren't pushing against air or the ground.

For a situation like this, two things must be true. First, the total vector momentum before the collision must be the same as the total vector momentum after the collision. Why do I say "vector"? The vector part is important because when one car is moving in the opposite direction as the other car, the total vector momentum is zero. After the cars crash and stop, the vector momentum will still be zero.

Second, the total energy before and after the collision must be the same. It is constant if there are no forces doing work on the system of cars plus the target. You have to include the target in the system so that you can account for all the objects involved in the interaction. Clearly, this would be true in space. Before the collision, all the energy is essentially kinetic energy. After the collision, this energy can be either in kinetic energy of the moving objects or in something I will call structural energy in the deformation of the vehicles.

Okay, now for a special case. Suppose two rubber cars move towards the target with the same initial speed. After the collision, they bounce back with the same speed. In this case, it is clear that both momentum and energy are conserved. Actually, the initial kinetic energy would be the same as the final kinetic energy. This means there must be no change in any other types of energy—so no structural changes and no damage. Really, this is sort of the boring case.

What if the two cars start with the same speed, collide into the target at the same time, and then stop? In this case, momentum is still conserved. Because they are moving in opposite directions, the initial momentum is zero; because they stopped, the final momentum is also zero.

What can we determine about the energy? Clearly, the kinetic energy is not the same before and after the collision because the cars are at rest after the collision. Notice the key difference in kinetic energy and momentum. Kinetic energy is a scalar quantity and always positive, meaning it doesn't cancel for two cars going in opposite directions. So, what happens to all of this kinetic energy? It goes into "damage" of the target. Simple, right?

Now, back to the MythBusters' collision strategy of using a stationary car and one moving at twice the speed.

In this case, the initial momentum is not zero. This means that even if the cars get as smashed as they can, they won't be at rest after the collision. If everything "sticks together" during the collision, it will still be moving off to the right afterwards because that is the direction of the original momentum.

What about the starting energy? If one car has twice the initial velocity and the other is at rest, it won't equal twice the total energy. Why? Because the kinetic energy depends on the velocity squared. This means that even though the other car is stationary, the moving car still has more energy than the case where both cars are moving. The problem is that all of this energy cannot go into "damage" because the cars can't be at rest after the collision when the initial momentum is not zero.

All of the previous discussion was in the special case of a space-based collision (for simplicity). What if the stationary car is held in place by a wall?

The external forces make a huge difference in the outcome of this scenario in comparison to the previous scenario which only had two cars plus the target as the system. In this scenario, the force of the ground on the stationary car is important—the initial and final momentums are not the same.

What does this scenario tell us about the energy? Even though there is an external force on the system, it doesn't add any energy because it does not move. Work is the product of the force and the distance the force moves. Without work, the total energy post-collision would be the same as the initial kinetic energy. Still, in this case the total energy would be twice the amount from the case with the two cars moving at the same speed.

In the end, doubling the speed for just one of the cars and keeping the other one stationary is not the same as two cars that are both moving. It seems like a good idea and an easier way to do the experiment, but it doesn't work.

Don't worry, *MythBusters* is still a great show. Mistakes like this point out that Adam and Jamie are just regular people trying to do awesome science.

BIG HAIL IS BAD

It seems every spring there is another crazy hail video posted on YouTube. If you want an example, check out this one from April 2012 in St. Louis: http://youtu.be/DDvuhENw3BE. With the abundance of cell phone cameras, along with the ability to share videos on YouTube, it is easier to see the effects of devastating weather events like this. The crazy thing in this case is that the hail gets big enough to break car windows.

Why is the large hail such a problem? It's probably obvious that larger hail has a larger mass. The less obvious problem is that larger hail also falls faster. Let's calculate how fast hail falls based on its size.

First, let's just assume all hail has the same density. Like all assumptions, this probably isn't strictly true, but it will be true enough for a nice estimate. If the hail is just ice, it could have a density somewhere around 917 kilograms per cubic meter. This is a lower density than water, which is 1,000 kilograms per cubic meter. It's useful to compare the density of ice to that of water since we know that ice floats as does everything with a density lower than water.

As the hail falls through the air, there are two forces acting on it. There is the gravitational force that pulls the hail down. The magnitude of this force is the product of the hail mass and the gravitational field (g). As I said before: the bigger the hail, the greater the mass. The mass depends on the volume and the density of the hail. For spherical hail, if you double the radius you would increase the mass by a factor of eight since the volume is proportional to the radius to the third power.

There is another force acting on falling hail, air resistance. The typical model for air resistance says this force depends on the shape of the object, the density of air, the cross-sectional area of the object, and the square of the speed that it is moving. If you double the radius for falling hail, you will increase the cross-sectional area by a factor of four since the area is proportional to the square of the radius. Perhaps you can see the problem already. Gravity pulls the hail down and air resistance pushes up on falling hail. By making the hail bigger, both the air resistance and the gravitational forces will also get larger. However, these two forces don't increase the same amount.

If the hail is allowed to fall for a sufficient amount of time, it will fall faster and faster. Of course, as it falls faster the air resistance force increases. Eventually it will reach terminal velocity. At terminal velocity, the air resistance force and the gravitational force have the same magnitude. This makes the net force on the hail zero so that it doesn't change speed. If I know the size (and thus the mass) of the hail, I can calculate this terminal speed.

Let's look at two hail sizes (pea-sized and baseball-sized) and compare the terminal velocities. If the pea has a radius of about 0.2 centimeters, then its terminal velocity will be around 10 meters per second (around 22 mph). Increasing to baseball-sized hail with a radius of about 3.5 cm would give a terminal speed of 40 m/s (90 mph). That's a big difference in speed.

Of course, it's not just the speed that matters with hail. When hail colides with stuff, there are two things we can look at: momentum and kinetic energy.

Which one is the best to consider? This is not such a simple question. Let's look at the kinetic energy first.

Since I already have the terminal velocity for both pea-sized and baseball-sized hail, all I need to do is to plug in these velocities, along with the mass, into the kinetic energy equation (1/2 mass times velocity squared). The pea-sized hail will only have 0.001 joules of kinetic energy. The baseball-sized hail is going faster and has a bigger mass. This puts its kinetic energy at 122 joules.

How could you get a grasp of these energies? What about a comparison with the kinetic energy of a bullet? A .22 caliber pistol bullet has about 100 joules of energy. Compare this to a .45 caliber bullet with a kinetic energy around 500 to 800 joules. Does this mean getting hit by baseball-sized hail is like getting shot with a .22 caliber bullet? No. Let's look at the momentum of the hail and then come back to comparisons.

The baseball-sized hail would have a momentum around 6 kg*m/s. If you compare this to bullets, a .45 caliber has a momentum of around 4.5 kg*m/s and a .22 is around 1 kg*m/s. In reality, this large hail is much more like a real baseball thrown from a major league pitcher.

What if I had a steel sphere with the same mass and size as the baseball-sized hail? Of course to do this it would have to be hollow. If I dropped the hail and the steel sphere, they would reach the same terminal speed and have both the same momentum and kinetic energy. However, what would happen if they hit your car windshield? They wouldn't do the same thing. Why is this? Mostly because the hail is more likely to deform during the collision than the steel would be. Here is a diagram showing the two spherical objects some short time after the initial contact (but before they stop):

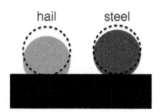

hail   steel

Since the ice of the hail gets compressed more than the steel ball, this means two things. First, more compression means more time. If the collision between the hail and the surface takes longer, it will exert a smaller force on the object. This is because of the nature of force and momentum. Essentially, the net force is equal to the time rate of change of momentum. Both the steel and ice balls have to decrease to zero momentum. If the steel ball does this in a shorter time, it will require a larger force. So, just because they have the same momentum or energy, doesn't mean they will impact in the same way.

This example with hail shows a common problem with size. We often think of big things acting just like smaller things, but that rarely happens. In the case of hail, the air drag and weight depend on different powers of the radius such that they don't cancel. Bigger hail will fall at a faster terminal speed and have more kinetic energy on impact. This is why big hail is bad. It's best to just stay out of hail and cover your car to prevent damage.

A SCALE AT THE BOTTOM OF A POOL

Here is a question submitted by a reader of my blog:

"An Olympic-sized swimming pool is filled with 660,000 US gallons of water. An imaginary scale under the pool reads 5,511,556 pounds—the weight of the water. Now a 12,000 pound, 5 foot wide spherical wrecking ball is lowered halfway into the water by a crane. What does the scale read?"

First, let me tell you what the wrong answer to this question is: 5,511,556 pounds. It's easy to see why some people would get to that answer. If the crane is still holding the wrecking ball, how could it add any weight to the scale? Wrong! Here's another wrong answer: 6,000 pounds. After all, half of that steel ball is not part of the total mass being weighed, right? Nope. Okay, one more wrong answer that sounds smarter: 5,511,556 pounds, but this time because of Newton. If the ball is pushing down, that force is canceled out by the water pushing up, so the scale doesn't change, see? Okay, closer, but still not right.

What happens when you lower a steel ball halfway in the water?
Here is a force diagram:

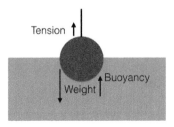

There are three forces on this ball. First, there is the tension. There must be a tension force to keep the ball up (a solid steel ball wouldn't float). Then there is the gravitational force (mg) where g is the gravitational field. But what about this F_B force? This is the buoyancy force. Essentially, it is the water pushing up on the ball.

What is the value of this buoyancy force? Well, suppose the ball was replaced with some water like this:

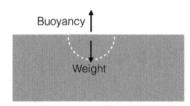

This diagram shows the water that would be there if it weren't displaced by the ball. What can I say about the forces on this water? Well, there isn't a string holding this one up so there are only two forces on this part of the water. There is the gravitational force pulling down on the water and the buoyancy force pushing up on the water. If I assume this part of the water is stationary, then these two forces must have the same magnitude.

Why is there even a buoyancy force anyway? One way to think about the buoyancy force is to consider the collisions from the water outside the object colliding with the object. Here is the cool thing: these collisions with the water outside are the same whether that object is a steel ball or some other water as long as the two have the same shape. This is great because I know what the buoyancy force on the chunk of water must be: it has to be the weight of that water. Since this is the same shape as the steel ball, then the buoyancy forces will be the same. This way, I can say the magnitude of the buoyancy force is equal to the density of the water, the volume of the object, and the gravitational constant (g).

Well, what does this have to do with a scale at the bottom of a pool? Newton's third law, that's what. First, let me publicly state that I really prefer to call Newton's third law "the definition of force." Basically, this is the idea where forces are an interaction between two things. If the water pushes up on the ball with a force F_B, then the ball has to push back down on the water with a force of the same magnitude.

Up to this point I have been looking at forces on the ball. Let me now pretend all this water is sitting on a scale that measures its weight. Here is a force diagram for the water before the ball is lowered into it:

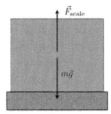

Yes, there is nothing holding the water. It is merely sitting on a scale (just for simplicity). Now I will lower the ball into the water. Since the water pushes up on the ball, the ball has to push down on the water.

Here is that force diagram:

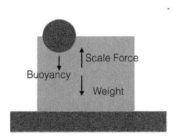

With this new force on the water, what happens? Well, the water is still stationary. This means the net force must be zero (the zero vector). If there is another force pushing down, how can the forces still add up to zero? The mass of the water doesn't change since nothing was added or taken away. The only thing that can change is the force the scale pushes up on the water. It has to increase which means the scale reading will increase. How much will it increase? By an amount equal to the weight of water displaced by the object.

There are two interesting points here. First, this change in scale reading doesn't depend on the material of the object in the water. It doesn't matter if the object is steel or balsa wood. If it displaces the same volume of water, it will change the scale reading the same amount. Oh sure, balsa wood wouldn't sink down that much. You would have to push it down.

The other thing to consider is the scale. From the scale's perspective, how would it seem like there should be more water to support? I know scales don't really think about problems like this. Normally, scales are more concerned with issues like being "zeroed" or making sure they are plugged in and locked down. But sometimes you get a scale that will consider issues like this. From the scale's perspective, there is more water to support. If I put a ball in the water that displaces a volume of 1 m³, then where does this displaced water go? This ball will cause the pool water level to rise by an amount of 1 cubic meter. So, at the bottom of the pool, it looks like there is more water (it's deeper).

The great thing about this pool-office question is that people tend to not believe the answers. Well, to help understand this problem I did a little experiment. I put a beaker of water on a scale. The mass of the beaker plus the water is 254 grams. Next, I lower a steel ball halfway into the water. In order to measure the tension needed to hold the ball up, I lowered the ball using a spring scale.

With the same amount of water and the ball halfway in, the scale reading increased from 254 grams to 268 grams. By the way, the mass of this ball is 206 grams. What if I replace the steel ball with a wooden one of the same size? From what I said before, the scale should change by the same amount.

The scale increases about the same amount (13 gram increase). See. I told you. Of course, this really just shows that you could have probably answered this question by yourself. It wouldn't be too difficult to find a scale and a glass of water. Just lower a ball into the water and see what happens to the scale.

It's still a great question with an even better answer.

PI AND GRAVITY

It's no secret that I love the number pi. Really, what's not to love? Now for a trick. Get out your calculator. What is pi squared? You can use the value for pi built into your calculator or you can just use 3.1415 or something with as many digits as you feel comfortable with.

Did you do it? Does the result look familiar? On my calculator, pi squared equals 9.869. This looks very close to the value of the local gravitational field near the surface of the Earth in newtons per kilogram. Doesn't it?

Wait, what is this local gravitational field? Isn't 9.8 the acceleration due to gravity in meters per second squared? Yes it is. That is what many people call it, but it isn't the most appropriate name. Let's look at a simple book placed on a table. The book is just sitting there motionless, right? That means it isn't changing its momentum. The momentum principle says the net force is equal to the rate of change of the momentum so the net force must also be zero. There are really just two forces acting on this book. There is the gravitational force pulling down and the force from the table pushing up.

What is the magnitude of this gravitational force? The common reply is that the gravitational force is the product of the book's mass and g. On the surface of the Earth, this is indeed true. But what is g? If you say it is the acceleration due to gravity, there is a slight problem. The book isn't accelerating, is it?

Now, if you drop the book, then the only force acting on it will be the gravitational force. In this case, it will accelerate downward at 9.8 meters per second. But since that is only one special case, it is better not to call g the acceleration due to gravity and instead call it the gravitational field.

What about the units? If g is the gravitational field, it should have units of force per mass just like the electric field has units of force per coulomb. Really, calling g the gravitational field isn't just the right thing to do, it helps introductory physics students grasp the concept of a field. When they get to the electric field in class, it won't be such a crazy idea.

Alright, back to pi squared. You might think it is a cosmological coincidence that pi squared is close to the value of g. It's not by chance that the values are close. But if it's not a coincidence, why isn't pi squared exactly g? The answer is that g is not exact. The value of g on the surface of the Earth depends on several things. First, there is the problem of apparent g versus real g. The apparent g is the value we measure on the surface of the Earth. It takes into account the acceleration of our reference frame due to the rotation of the Earth. The closer the observation location is to the equator, the lower the apparent value of g.

The other factor influencing g is the non-uniform nature of the Earth. If you are closer to a region with high density rocks, the value of g will increase. This means that as you move around from place to place, g will also change. There is not just one value for g.

Now for something even more interesting: why does this pi-g relationship even exist? It has to do with the definition of the meter. Before that, let's look at the seconds pendulum. This is a pendulum that takes exactly one second to go from one side of its motion to the other (or a two-second period). I bet you have seen a seconds pendulum before. A grandfather clock has a period of two seconds. Boom, there is your seconds pendulum.

Now, technically a grandfather clock isn't a seconds pendulum. A seconds pendulum is a point mass on the end of a massless string (you can get massless string at your local hardware store). The grandfather clock, on the other hand, has a stiff arm that swings back and forth. Thus it has a center of mass that is not the same as the length of the swinging arm. If you measure the length of the arm in a grandfather clock, it will still be pretty close to 1 meter. An actual seconds pendulum has a length of essentially 1 meter. Of course, if you made one and your friend on the other side of the world made one, they might be different lengths.

Go ahead and try it. Get a small mass like a nut or metal ball. Metal works well since its weight will likely be significantly larger than the air drag force, enough so that you can ignore it. Now, make the distance from the center of the mass to the pivot point 1 meter and let it oscillate with a small angle (maybe about 10°). If you like, you can make a video or just use a stopwatch. Either way, it should take about one second to go from one side to the other.

I'm not going to derive it, but it isn't too difficult to show experimentally that for a pendulum with a small angle the period of oscillation is:

$$ T = 2\pi \sqrt{\frac{L}{g}} $$

If you want a period of two seconds, you can solve for the length. Doing that gives you a length of g over pi squared. If you define this length as 1 meter, as was the original definition, you get a value for g of 9.8 meters per second squared. Wait! What about g in newtons per kilogram? Well, the two units are equivalent, so there.

But why is pi in the period expression for a pendulum? That's a great question. Is it because the pendulum moves in a path that follows a circle? No. The equation of motion for an oscillating mass on a spring (simple harmonic motion) has the same form as the small-angle pendulum and it isn't moving in a circle. Then why is pi in the equation? I guess the best answer is that the solution to simple harmonic motion is a sine or cosine function. I don't know what else to say other than it gives us a solution. Because we have a sine function for the answer, the period would have to have a pi in it.

It's really surprising how many different situations connect things we didn't think were connected. The number pi and the gravitational field are a great example of this. Of course, you might be surprised to find the number pi can pop up in other situations that don't involve circles.

GRAVITY AND THE MASS OF ALL THE HUMANS

This was another great question (reworded for clarity):

> *"I understand that the gravitational force that the Earth*
> *pulls on the Moon depends on the mass of the Earth.*
> *So, as the human population increases does this mean*
> *that we will eventually pull the Moon closer to the Earth?"*

Let's look at this in different parts. First, what is the mass of all the people on the Earth? As I am writing this, there are about seven billion humans living on the Earth. Hopefully when this book is finished, there will still be around seven billion people, but let's wait and see.

If I know the average mass of a person, I can get the mass of all the people. Here's where we guess. Well, not a completely random guess. No, it will be a well-aimed guess. An average human male is maybe 160 pounds (70 kg) and a female is about 110 pounds (50 kg). I suspect those are a little high for all of the adults. I would guess that adults in the United States are a bit more massive than other parts of the world. Now, what about children? I could say there are about as many men as women. That would put the average mass somewhere around 60 kg. If I take into account children, maybe the average mass is around 40 kg.

Yes, 40 kg is an estimate. However, it isn't a crazy estimate. If I were to actually go around and put every human on a scale, what would I get? I'd get a broken scale, that's what. But I could also get the exact average human mass. I don't think this mass could be lower than maybe 20 kg. Also, it couldn't be higher than 60 kg (unless we discovered a new race of giant humans somewhere). So, again, 40 kg isn't a crazy average. A crazy average would be something like 100 kg, that's crazy.

Now that I have an estimate for the average mass of a human, the total human population mass is just going to be the total number of humans multiplied by the average human mass. This would give a mass of 280 billion kg (2.8 x 10^9 kg). That's a large mass, but how does it compare to the mass of the Earth? At about 6 x 10^{24} kg, all the humans are just a tiny fraction of the total mass (4.7 x 10^{-14} %).

It's hard to realize just how small a percentage of mass humans are. Let me give another example. What if we look at the mass of a 60 kg human? The mass of all the humans on the Earth is proportionally like the mass of one single yeast cell on a human. So as you can see, the mass of the human population just doesn't matter in regards to the total mass of the Earth.

However, there is another part to this question: does the mass of the stuff on the Earth change? By "stuff" I mean all the living things and consumable things like air and water. In short, the answer is no. Where do humans come from? Or perhaps I should ask where the mass that makes up humans comes from? As a person grows or as new people are created, the mass for this new material comes from three sources: air, water, and food. Actually, I'm not too sure how much of our material comes from the air but it could be a small contributing factor.

Where does the food come from? If everyone was a vegetarian, the food would come from plants. But where does the mass of plants come from? Most of the plant mass comes from the air. Yes, the air. Plants take in carbon dioxide and produce oxygen. They save the carbon (and water and other stuff) to use as building blocks for growing. It might seem strange, but it's true.

Which means that, indirectly at least, the mass of the human population comes from the air. When people die, they decompose and produce more carbon dioxide. It's an endless cycle. Just about all of the mass comes from something that is already on the Earth. Yes, I said "just about."

Is there mass that leaves the Earth? Is there mass that is added to the Earth? The answer for both questions is yes.

When does the Earth lose mass? First, there is loss of gas in the atmosphere. Think of our air as a bunch of gas particles bouncing around. That's essentially what is happening in reality. Some of these particles of gas (oxygen or nitrogen molecules) are going quite a bit faster than other particles. If the particle is going fast enough and is near the top of the atmosphere, it can escape the Earth's gravitational influence. This does happen, but the effect is quite small. The other way for the Earth to lose mass is when humans send objects into space. Again, the total mass of all the man-made space objects is quite small.

The Earth also gains mass. According to NASA, there are about a hundred tons of meteoroids hitting the Earth each day. Over a year, this would be 3×10^7 kg. It would take about a hundred years' worth of meteors to equal the mass of the human population. However, doubling something super small is still something super small.

In the end, the mass of the Earth does change, but not from the human population. This mass is still relatively small compared to the mass of the Earth and doesn't make a significant contribution to the Earth's gravitational field. Even if the mass of the humans was much greater than it is now, the overall mass of the Earth is still mostly constant.

WHY DO MIRRORS REVERSE LEFT AND RIGHT, BUT NOT UP AND DOWN?

At some point all of my children have noticed that when you look in a mirror, some things look backwards. Yes, when you look at yourself in a mirror, your left hand appears to be on your right side. Mirrors seem to reverse left and right. However, your head doesn't appear to be where your feet should be? So why doesn't a mirror reverse up and down too? Here is the short answer:

ANSWER: Mirrors don't reverse left and right and they don't reverse up and down. Wouldn't it be kind of funny if I just stopped here? But you know I can't.

First, go and stand in front of a mirror.

What should you see here? The image of the top of your head is at the top of the mirror. The image of your right hand is on the right side of the mirror. The image of your left hand is on the left side. There is no switching. Why is there a problem?

I think the problem is that if you were on the other side of the mirror (and not just an image), your right hand would be on the left side of the mirror and your left hand would be on the right. You can see videos on YouTube where identical twins have played this trick for television shows. Two identical rooms are set with glass in between, and an identical twin on each side. The prank is that other people coming into the room can't see their reflection in the mirror, but they look to the identical twins and see what they think is a reflection. In these videos, when one twin raises their right hand, the identical mirror twin raises their left hand. So we think the mirror reverses left and right.

Another source of confusion might come from the same reason why this seems like a physics question when it is really not. Many of us know that the lenses in a microscope, telescope, and even cameras flip an image. This can make a telescope confusing to get into position for novices. Binoculars flip the flipped image back, because no one would want to watch a baseball game or a bird upside down. In fact, when light passes through the lenses in your eyes, it is also flipped upside down on the back of your eyeball. But that's how our brain is programmed to process it. Why does this happen? Basically, light coming down from the sky keeps traveling down as it passes through the focal point of the lens in your eye, hitting the bottom of the back of your eye. Light coming up from the ground keeps traveling up as it passes through the lens and ends up at the top of the back of your eye.

Nothing like this is happening with the mirror. It is just reflecting your image directly back to you. If you walked around to the back of the mirror, your left would be on your right, and your right on your left. However, your head would still be at the top of the mirror. Well, this is only true if you walked around to the other side. But what if you jumped over it and landed upside down? Now your head would be across from your feet, and your feet across from you head. But your right hand would be on the right side of the mirror and your left on the left. So, from this perspective, it did reverse top to bottom instead of left to right. Mind blown? I'm not done yet.

Now we can see the problem. The problem of the mirror comes from our cultural background. We assume the image is like someone walking around to the back side of a mirror. If we assumed people would get to the back side by going head first over the top, I would be answering a different question. That question would be: why do mirrors reverse up and down?

There is something that mirrors do flip. They flip front and back. Just imagine that you could walk forward into the mirror world. Your right hand would still be on the right side. And your head would still be at the top. However, your back would be facing towards where the mirror was, not your front. So, mirrors flip front and back.

CHAPTER 2: SUPERHEROES

WOULD THE HULK BREAK UP THE ROAD WHENEVER HE JUMPED?

The Hulk cannot fly, but from comic books to the hit *Avengers* movie, he sure can jump high. What kind of forces would the Hulk exert on the ground when he jumps? Before guessing at some of the values, let me start with the general case of a jumping Hulk. Consider the Hulk in three positions during the jump.

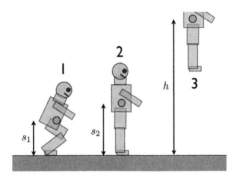

I didn't draw the forces, but during the jumping part there would be the gravitational force and the force of the ground pushing up. The best approach in a problem like this is the work-energy principle. The work-energy principle works best when dealing with changes in position instead of changes in time.

What does the work-energy principle say? It says that the work done on a system is equal to the change in energy of that system. What is work? In its simplest form, work is a force applied over a distance (yes, it is technically a little more difficult than that, but I am trying to keep it simple). What is energy? If I include the Hulk plus the Earth as the system, the system can have kinetic energy (energy for things that move) and gravitational potential energy.

Let me just point out something about the work and energy. They are usually defined as follows: "energy is the ability to do work" and "work is something that changes energy." Yes, this is a rather circular definition. Why? Because the work-energy principle is really just a frame of reference we use to calculate things. It isn't necessarily something concrete. It just turns out that for every system we look at, the work-energy principle is valid.

Let's get back to the Hulk. In order to look at the important part of the jump, we have to work backwards. Going from position two to three, the Hulk is just like plain projectile motion in that his motion is only based on the gravitational force. He isn't doing anything but moving up and slowing down. When he gets to his maximum height, he stops for a moment. This means that at the highest point, total energy for the Hulk is just in gravitational potential. By measuring the height of his jump, I can calculate the work needed to be done during the jumping part (position one to two). Also, with the distance the jump occurs (while in contact with the ground), I can find the force the ground exerts on the Hulk.

Now I need some estimates to use in this calculation. First, I need the mass of the Hulk. This is a tough one. The Hulk is shown in so many different ways that he can have a huge range of values for his mass. Let me assume he is the same density as a normal human and a size as shown in the *Avengers* movie.

In one shot, the Hulk is shown standing next to Hawkeye. If I assume Hawkeye is a normal-sized human (around 1.8 meters tall), then the Hulk would be about 2.5 meters tall. This is basically a guess since the Hulk is sort of bent over, but I am going with it. But what about the mass? The Hulk isn't just taller than Hawkeye, he's thicker and wider. Suppose a human and the Hulk were both cylinders. In this case, I could draw the following:

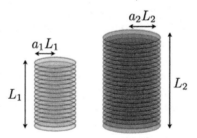

Here, I am assuming there is some relationship between the height of a person and the radius of a cylinder representing this person. I have the constant "a" that relates these two values. Let me say that the Hulk isn't a normal human. He is bigger, but also bulkier. I will approximate his radius to height ratio as 1.25 times larger than Hawkeye's (just a guess). So, if Hawkeye has a mass of 70 kg and the Hulk has the same density he would have a mass of 293 kg (645 pounds).

Bam! That is a huge mass. The important thing to remember is that just because the Hulk is 40 percent taller than a normal human doesn't mean his mass is 40 percent larger.

I feel like I should say something about these cylinders. I can hear the complaints now: "Hey! People aren't cylinders!" This is, of course, true. However, it isn't a terrible thing to approximate a person as a cylinder. It gives a fairly nice estimate for these things. This reminds me of the famous spherical cow. If you hang around physicists for a significant amount of time, someone is going to bring up the spherical cow. What is this thing?

I guess I can give a quick explanation. Suppose I am in a room and I throw a pencil. Why am I throwing a pencil? No worries, I sometimes do random things. However, suppose I also want to model the motion of this projectile pencil. What do I do? Some things I could consider:

- Should I consider it to be a flexible object?
  If so, when I throw it will oscillate and flex.

- As a semi-rigid object, I still need to consider it's rotation
  in three different directions. This turns out to be non-trivial.

- What about air resistance?

- What about variations in the density of the air in the room?

- Is the gravitational field in the room constant?
  What about the gravitational interaction between
  the pencil and that massive desk over there?

- Does the pencil have any excess charge that would produce
  and electrostatic interaction with other objects or an
  electromagnetic interaction with the Earth's magnetic field?

That should be enough for you to get the idea that this could be a complicated problem. But what if I just want to get a value for the initial velocity based on the horizontal distance it travels? In that case I could just ignore the above factors. Yes, it would be wrong, but I could get a pretty good answer. In reality, what I would be doing is reducing the complicated problem to the following:

- A point mass.

- Only the gravitational interaction
  with a constant gravitational field.

This makes the problem quite doable. In physics (and science) we try to make models. They don't have to be perfect models. In fact, they never are. We just want models that work. If I assume the pencil is a point mass, that model works pretty well.

Back to the spherical cow. I don't know the origins of this joke, but there is one I always remember.

> *There is this dairy with cows and everything.*
> *The dairy farmer wants to increase her production of milk.*
> *To do this, she hires three consultants: an engineer,*
> *a psychologist, and a physicist.*
>
> *After a week, the engineer comes back with a report.*
> *He says: "If you want to increase milk production,*
> *you need to get bigger milk pumps and bigger tubes*
> *to suck the milk through."*
>
> *Next comes the psychologist.*
> *He says: "You need to make the cows produce more milk.*
> *One way to do this is to make them calm and happy.*
> *Happy cows produce happy milk. Paint the milking stalls green.*
> *This will make the cows think of grass and happy fields.*
> *They will be happy."*
>
> *Finally, the physicist comes to present her ideas.*
> *She says: "Assume the cow is a sphere . . . "*

Maybe now the joke can make sense to non-physicists.
Either way, I still think it's funny.

How did I get off track? Let me go back to the mass of the Hulk. There is something that always bothered me. Bruce Banner is a pretty normal-looking human, right? But then he turns into the Hulk. So, if he goes from 70 kilograms as a human to almost 300 kg as the Hulk, where does the extra mass come from? What if this is conversion of energy to mass from Einstein's $E = mc^2$? This would take 2.7×10^{19} joules of energy. Where does that come from? The total power output from the Sun is about 4×10^{26} watts. However, only about 1.7×10^{17} watts hits the Earth. If the Hulk used all of this solar energy, it would take over two and a half minutes in order to capture enough energy to "transform." I guess this could be the "getting angry time."

But what if the Hulk doesn't change mass? In this case, he would still be 70 kg but have a different density. Solving for the density, he would be 0.24 times the density of a human. A good starting estimate for human density is to say they are similar to water, which has a density of 1,000 kg/m³. This would put the Hulk's density at 240 kg/m³. Just to compare, this similar to the density of cork. Crazy.

What's next? I need to estimate the height of the Hulk's jump. If you look at the trailer for *The Avengers* it seems they used models of real buildings. Using Google Earth, I played around until I found what might be the exact location in the movie where the Hulk makes one of his jumps. From the map and the height of nearby buildings, it seems like he jumps to a height of about 400 feet (120 meters).

Now, we just need to plug in my values. I get an average force of 4.08×10^5 newtons. That is how hard the Hulk pushes on the ground and how hard the ground pushes on the Hulk. Yes, this is an average force. But this is also the smallest overall force. If I consider a non-constant force, that means there will be some part of the jump with a lower force but also some part with a greater force. I want to use the case with the best possible jump with the lowest overall value of the force.

Would the Hulk break the concrete? This is really the question I was aiming for. My suspicion is that during a jump, the Hulk would push so hard on the concrete that it would crack. How can I find out if the concrete (or whatever the surface may be) survives? I need to look at the compressive strength. This is the maximum pressure a material can withstand before failing.

With some conservative estimates for the size of the Hulk's feet, I get a pressure of 2.9 megapascals. According to The Engineering Toolbox[1] , concrete has a compressive strength around 10 MPa. Well, I guess he might not crack the road during this type of jump. Or would he? I suspect the force versus displacement curve for a jumping Hulk is not constant and has a peak in it.

If the peak goes over the force that would produce a larger pressure, the road would crack. Also, in my calculations I used the total area of the Hulk's feet. This would be the correct thing to use if he jumped flat-footed. People don't normally jump this way. Instead they end up pushing on the balls of their feet. This would decrease the area of contact and increase the pressure.

So, in the end, I suspect the Hulk leaves cracks in the road whenever he jumps, but he's not tearing the street apart.

---

1   http://www.engineeringtoolbox.com/compression-tension-strength-d_1352.html

THE PHYSICS OF THOR'S HAMMER

I was introduced to Thor during my teenage years. It was then that I spent a lot of time with comic books. I hate to show my bias, but I was mostly into Marvel comics and rarely picked up a publication from DC comics. In the Marvel Universe, Thor wasn't my favorite superhero, but he was pretty cool.

Oh? You haven't heard of Thor before? You didn't even see The Avengers? Okay, let me get you up to speed. In the Marvel Universe, Thor is a Norse god who was banished temporarily to the Earth. Or maybe he is an alien, I can't remember how the movie made things work. The point is that he has some super powers. He also has a super hammer named Mjölnir. I won't write the name of his hammer too much because it is a pain in terms of special characters.

In one version of the hammer's origin, the god Odin orders dwarven black-smiths to forge the hammer from the core of a star. Note that even in the Marvel Universe, there are multiple origin stories of Mjölnir. But if I stick with the core of the star version, I can calculate the mass of this hammer. It makes the most sense for what I am going to do.

If you take some material from the core of a star, what would it be like? Would it be hot? Yes, it would be extremely hot. Would it have a high density? I guess so. But this is a great time to talk about stars.

In short, a star is just like a planet. Well, it is like a planet in that both planets and stars are accumulations of materials. Suppose you have a giant cloud of hydrogen gas in space. Where did the hydrogen come from? Let's just say it is there for now. Since all of the hydrogen atoms have mass, they all exert a grav-itational force on each other. This gravitational force is quite small, but over time it can cause the hydrogen cloud to condense.

Typically, this collapsing gas cloud can form a solar system (a star with plan-ets), but let's just look at the star. If this giant collection of hydrogen gas has formed a sphere like a star, what also has to happen? Why doesn't it keep on shrinking? What stops this collapsing process? I guess the best answer is the same answer that keeps the oceans from collapsing into a thin layer of water at the bottom of the ocean.

Water particles in the ocean are supported by collisions with other water parti-cles that are beneath them. As you go deeper and deeper in the ocean, there needs to be more and more collisions from the still lower water particles in order to support everything above. This means at deeper depths, the pressure in the water has to go up. If it didn't, the whole ocean would shrink down to some crazy high-density, thin layer of water at the bottom.

Essentially the same thing happens in a star. The two big differences between a star and the ocean are that stars are much bigger and they are not made of water. Liquid water has an interesting property that makes its density mostly constant in large bodies of water. If you take a hydrogen gas and increase the pressure, the density would also increase. An ocean can be a few miles deep. Even a smaller star like our sun has a radius of around 400 thousand miles. The large size means super high internal pressure to prevent it from collapsing. Along with this pressure comes a very high density.

Suppose Mjölnir was created from the material found in the center of our neighborhood star, the Sun. The density of this material is around 150 grams per cubic centimeter. That's pretty high. Recall that water has a density of one gram per cubic centimeter and lead is 11.4 grams per cubic centimeter. It's not just very dense, it is also very hot at around 15 million kelvins (a light bulb filament is around three thousand kelvins).

What kind of material is this very dense and very hot material? It would mostly be free protons and ionized helium. At this temperature, you wouldn't have hydrogen atoms. The electrons have too much energy to stay bound to one particular atom. Instead, they are free to zoom around. The same is true for the helium. It would just be two protons and two neutrons bound together but without the electrons.

How would you get this material out of the core? I have no idea. If you did get it out, it would obviously be hot. This is good in the sense you could form it into whatever you wish. Maybe you would like to make it into a hammer. But there are a couple of problems. First, this proto-hammer would be so hot that anything near it would melt. That's bad. But the real problem comes when you cool it off. Protons don't like to stick around next to other protons since they both have the same positive charge. In the core of the Sun, they don't really have a choice. They are pushed together by the 400,000 miles of material above them. However, once you take this material out of the core, the protons and helium atoms would just start shooting off the lump of material. In short, the stuff would evaporate.

You can't keep helium around as a solid material, the same is true for hydrogen. In the end you would just have a nice, big surrounding area that was melted by this core material.

Could you make a hammer from another type of star? In larger mass stars, the hydrogen and helium can go through nuclear fusion to form heavier elements all the way up to iron atoms. Iron can exist as a solid at room temperatures, so it would be nice choice. But there are a couple of things to think about. First, if the star has a larger mass, the interior core will also have a higher density. Just how high depends on the size of the star. When a star is in the process of producing iron atoms, it can have a density in the core as high as 100 million grams per cubic centimeter. That is crazy. Also, the temperature would be around 2 billion kelvins.

Of course, as this material cools off (which would take quite a long time) it would expand. Once it got back to room temperature it should have a density of iron that we find here on the Earth with a value of about 7.8 grams per cubic centimeter. In the end, you would have just a normal iron hammer.

Normal is boring. What would happen if you had a completed hammer made with the core material and having a density of 100 million grams per cubic centimeter? It would have a fairly high mass. Suppose the hammer is a rectangular cube with dimensions of about 15 x 15 x 8 centimeters (not including the handle). This would have a volume of 1,800 cm³. Since density is mass divided by volume, the mass would be the density multiplied by the volume. This gives a mass of 1.8×10^{11} grams or 400 million pounds (on the surface of the Earth). Good luck picking that thing up.

What if Thor held the hammer 50 centimeters over your head on the surface of the Earth? In this case, there would be two gravitational forces acting on you. The Earth would be pulling down on you (for me, this would be about 160 pounds) and the gravitational force from the hammer would be pulling up since it also has mass. Normally, we would ignore these other gravitational forces.

Using the universal gravitational model, the hammer would pull up with a force of 0.33 pounds. Not incredibly high, but you would probably feel that. Of course, the closer you got to the hammer, the greater the gravitational force from it. If you put an object 3.5 centimeters from the center of the hammer, this gravitational force from the hammer would be the same as the gravitational force from the Earth. Too bad the hammer is thicker than 3.5 centimeters.

I guess this enormous mass could explain why no one can lift it, except for those deemed worthy.

Alright, enough about the composition of the hammer. There is one other cool thing. How does Thor fly? At one point, I incorrectly claimed Thor could even fly. Apparently, that is wrong and he doesn't fly. Instead, Thor throws Mjölnir but then holds onto the handle and gets pulled along. How can this happen?

Let's start with a simple model. Suppose Mjölnir and Thor are two objects with the same size and mass. If they were right next to each other, the center of mass would be right in the middle. Now, what if Thor throws the hammer? To do this, he would exert force on the hammer over time and the hammer would increase in momentum. However, forces are always an interaction between two objects. This means that whatever force Thor exerts on the hammer, the hammer exerts back on him in the opposite direction and for the same time. So when Thor throws the hammer, the hammer's momentum increases and Thor's momentum increases in the opposite direction.

But now what if Thor grabs the hammer after throwing it? Essentially the exact same thing happens. By grabbing the thrown hammer, Thor will exert a force on it and the hammer will exert a force on him. So, he could get himself to move by throwing, but he would get right back where he was when he grabbed it. Not very productive for "flight." Oh, and it doesn't even matter in this scenario if the hammer has a larger mass.

In general, we call this "conservation of momentum." If we have a system consisting of two objects with no external forces, the momentum of the center of mass will not change. If one mass has momentum in one direction, the other will be in the opposite direction to make the total momentum unchanged.

However, all is not lost. There is a way to get the total momentum to change. The answer is with the external forces. If Thor throws the hammer straight up, there will be an external force on the Thor-hammer system: the ground. This would indeed throw Thor and the hammer up. The same thing could be done horizontally with the frictional force. So the hammer can, in a sense, help Thor move through the air—but not fly. Unfortunately, this would require the same force and strength as it would take to jump. It would be like jumping with your arms instead of your legs.

There is one other problem. How do you turn while in the air? Suppose Thor was able to throw the hammer to get him in the air. Turning would require some other external force in order to change his momentum. The only way he could complete an in-air turn would be to throw the hammer and not hang on to it. If he threw the hammer to the right, this would push him to the left. Momentum would be conserved but he would also lose his hammer.

I hear your complaints already: "Mjölnir also has the power to return to its wielder." So, it must also have some external force on it. I don't know how this works. All I can do is apply known physics models like forces and momentum to made-up things like Thor's hammer. But it's still fun to think about these things.

CAPTAIN AMERICA PHYSICS

Let's look at Captain America. Just to be clear, I am only going to look at the movie version of Captain America from *Captain America 2: Winter Soldier* and not anything from the comic book version of Captain America. The two worlds (comic books and movies) don't always agree.

In one scene Captain America throws his shield at the Winter Soldier (because that's what Captain America does). But wait, the Winter Soldier just catches the shield and throws it right back at Captain America. The cool part is what happens when he catches his shield. The impact is strong enough to push him back a little bit. Is this enough to get an estimate for the mass of the shield? I think so.

In this estimation there are two parts. In the first part, I need to find out how fast the shield was moving before colliding with Captain America. After that, I can look at the interaction of the shield and Captain America as a normal physics collision. This leads to the second part where I need to find the recoil speed of Captain America plus the shield.

Since the order of calculations doesn't really matter, let's look at the recoil speed of Captain America after the collision.

How do you measure recoil speed from a video? The easiest way is to use some type of video analysis tool and determine the position of Captain America in each frame of the video. From the position and time data, you can easily find the recoil speed. But that won't work in this case. Why not? The reason is that the video doesn't easily present that data. In an ideal video, the scene would show Captain America along with some object that could be used to determine the size of things. On top of that, all of the motion would be perpendicular to the field of view of the camera. Motions towards and away from the camera are problematic because of changes in size of objects due to perspective. Unfortunately, this video does not offer a good viewing angle for analysis.

What about another method to determine the recoil speed? This can also be calculated by assuming Captain America starts his slide with some initial speed (the recoil speed) and then slows down with a constant acceleration which I can determine from the coefficient of friction.

If I estimate a coefficient of friction with a value of 0.3 (which seems reasonable for gravel on a hard surface) and a sliding time of 1.08 seconds from the video, this gives a recoil velocity of 3.24 m/s. Remember, this is the velocity of Captain America and the shield after he catches it.

In order to find the mass of the shield I need two more things. First, I need the mass of Captain America. This should be pretty easy to estimate since he is just a human (yes, a perfect human). Let's say he has a weight of 220 pounds or 100 kg. Now, what about the impact speed of the shield? I am going to have to get this value from the video.

There is a quick shot showing the shield right after it is thrown by the Winter Soldier. According to Wikipedia, the shield has a diameter of 0.76 meters. I can use this size to scale the video and get a plot of position versus time for the shield. From this, I get a shield speed of 19.5 m/s (43.6 mph). That's pretty fast for a shield, but I guess it's okay since we are dealing with superheroes.

How can I use all of this to find the mass of the shield? It's all about collisions and the nature of forces. When the shield interacts with Captain America, it pushes on him and he pushes back on the shield with the same magnitude of force. Why are these forces the same? The answer is simply that forces are always an interaction between two things. When you push on a wall, it pushes back on you with the same force. It is an interaction between you and the wall or an interaction between Captain America and the shield. Also, there is something else true about this collision force. The time that the force pushes on Captain America is the exact same time that Captain America pushes back on the shield.

In order to use these ideas, we first need to look at the momentum principle. This says when a net force acts on an object, it changes that object's momentum. Since the force on Captain America and the shield have the same value (but opposite directions), the change in momentum of Captain America and the change in momentum of the shield will also be opposite of each other. It's fairly straightforward to see that this is equivalent to saying the momentum of the shield plus Captain America before the impact is the same as the momentum of these two after the impact. We call this conservation of momentum (since it's the same before and after).

In this particular collision, Captain America and the shield move together as one object after the impact. We call this an inelastic collision, and it's pretty easy to calculate since the two objects have the same velocity afterwards. It's even better in this case since, before the collision, only the shield is moving. This means the momentum of the shield before the collision is equal to the momentum afterwards.

Now we are ready to use some of our values. We already estimated the velocity of the shield plus Captain America after the collision, and we know the velocity of the shield before the collision. Putting this all together, I can solve for the one thing I don't know: the mass of the shield.

Are you ready for the answer? With a recoil speed of 3.24 m/s and a shield speed of 19.5 m/s, the shield must have a mass of 19.9 kg (assuming a 100 kg mass for Captain America). That's a pretty massive 43.9 pound shield. With this we can also get an estimate for the density of the shield. If I assume it's a flat disc with a diameter of 0.76 meters and a thickness around 0.5 to 1.0 cm (just guessing) this would put the density range from 8,767 kg/m^3 to 4,383 kg/m^3. That seems reasonable. Iron has a density of around 7,800 kg/m^3 and titanium is around 4,500 kg/m^3.

It's still a pretty heavy shield. It would definitely take someone in good physical shape to actually throw this massive object. It would be much more difficult to throw the shield than it would be to throw a baseball or football. I guess that's why Captain America is a superhero.

COULD SUPERMAN PUNCH SOMEONE INTO SPACE?

Superman is so strong that he can do anything, right? Could he punch someone so hard that they ended up in space? Let's do this calculation with some estimations.

When I say space, you might say "outer space." But how high is that? The Earth's atmosphere doesn't just stop at some height. No, instead the density of air gets lower and lower until you can't even really detect it. But for this problem, we have to pick a height we can refer to as "space." I am going to pick 420 km above the surface of the Earth as "space." Why that value? Why not? That is about the height of the International Space Station's orbit, so I think it's a good choice.

How fast would a person have to be traveling straight up so that he or she would make it all the way to space? Just to be clear, this is after Superman has already punched the person. Of course if Superman punched a normal human, bad things would happen to that person's body. To avoid these problems, let's say Superman is punching a clone of himself. I will call this clone "Superman-b." So, while the Superman-b is moving up, there are only two forces acting on him. There is the gravitational force pulling down which decreases a little bit as he gets higher. There is also an air resistance force on Superman-b. As a first approximation, let's just assume there is no air resistance. Maybe all the air was sucked away from the Earth in a previous battle.

If we are looking at what happens over some change in position (from the surface of the Earth to 420 km high), then we should use the work-energy principle. This says that any work done on the system is equal to the change in energy of the system. If I consider the victim (Superman-b) plus the Earth as the system then there is nothing outside of the system to do any work (remember, this is after the hit from Superman). That just leaves the change in energy. There are two forms of energy in this case. We have the kinetic energy of the system and the gravitational potential energy.

We don't know the initial velocity of the person, but we do know the final velocity. At the highest point of this motion, the person will temporarily stop before falling back down. This means the final kinetic energy is zero. What about the change in gravitational potential energy? The gravitational potential energy for two interacting objects is inversely proportional to the distance between the centers of these objects. It also depends on the masses of the two objects and the universal gravitational constant. Putting this all together, I can solve for the one thing I don't know: the initial velocity of the person.

$$W = \Delta K + \Delta U_G = 0$$
$$0 = 0 - \frac{1}{2}m_h v_1^2 - G\frac{m_h M_E}{R_E + h} + G\frac{m_h M_E}{R_E}$$
$$v_1^2 = GM_E\left(\frac{1}{R_E} - \frac{1}{R_E + h}\right)$$

If I plug in what I know, I get a "launch" speed of 2,778 m/s (6,214 mph). Yes, that is fast. But Superman-b would have to be going even faster than that. Why? Air resistance, that's why.

Here is a diagram of Superman-b shortly after he was hit by Superman:

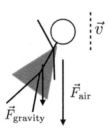

I will use the two following models for the magnitude of the gravitational force and the air resistance force:

$$F_{\text{gravity}} = G\frac{M_E m_s}{r^2}$$
$$F_{\text{air}} = \frac{1}{2}\rho A C v^2$$

For the gravitational force, the two masses are the mass of the Earth and the mass of Superman-b, while r is the distance between Superman-b and the center of the Earth. This force will decrease somewhat as Superman-b rises to space.

In the model for air resistance, A is the cross-sectional area of the object and C is the drag coefficient that depends on the shape of the object. The ρ is the density of the air. As you get higher in the atmosphere, this will decrease. As you can see, this air resistance force changes with both the speed and the altitude. Actually, the drag coefficient can depend on speed too but I will pretend like it is constant. So, this isn't such an easy problem.

Let me get estimates for some of these values. I am going to assume Superman-b is the same size and shape as a normal human. Maybe he has a mass of 70 kg. For the product of AC, let me estimate this based on the terminal speed of a skydiver. If a skydiver falls at 120 mph (54 m/s) then the air resistance would be equal to the weight of the skydiver. This means that AC would be 0.392 m². I will use an AC value of just 0.05 m². Why? Because the previous calculation was for a skydiver in a typical skydiver position. If Superman-b is "launched" in a head-up position, he will have a much lower cross-sectional area. This is probably way too low, but that's okay.

The other problem is dealing with a changing density of air. Fortunately, I have looked at air resistance at high altitudes before. The Red Bull Stratos Space Jump started at a point where the density of air was much lower than it is on the surface of the Earth. In the calculation of his falling speed, I used this model for the density of air.

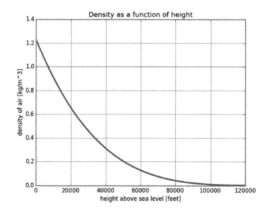

That model isn't truly valid for super-high altitudes. So, I will just use it for around 100 km and then assume the density of air is negligible after that.

Yes, I know it's wrong, but it will still work. First, I am trying to show that the starting speed of Superman-b is very large. Cutting off the density of air at high altitudes will give a smaller value of the starting speed. Also, when Superman-b reaches these high altitudes, he won't be going so fast such that the air resistance force will be small, even if there was some air up there.

What now? I can't directly calculate the required starting speed. However, I can pick a starting speed and create a numerical model to determine how high Superman-b will go. Then I can keep increasing the starting speed until I get the height that I want. For each starting speed, I will break the motion into tiny steps of time. During each of these steps, I will do the following:

- Calculate the density of air based on the height.

- Calculate the sum of the gravitational and air resistance forces using the height, density of air, and the speed.

- With this net force, calculate the change in momentum during this time step.

- Based on the momentum, determine the change in height during this time step.

- Repeat the above.

It looks complicated, but it isn't too bad. Using this model, a punched Superman-b with an initial speed of 2,778 m/s would only get to an altitude of about 6,500 meters. This is not even close to the required 420 km to be classified as "space."

What if I just keep running the model with air resistance associated with higher and higher speeds? Eventually the speed will be high enough to reach outer space. Here is a plot of the maximum altitude as a function of starting speeds up to a speed of 10^5 m/s:

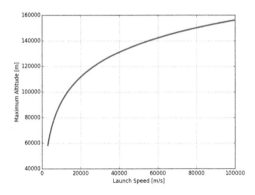

Even at 10^5 m/s, Superman-b would only get to a height around 13 km. I am a little disappointed. I thought I would get Superman-b higher than that. Even if Superman punched from the top of Mount Everest, it would still be tough to get Superman-b into space. The problem is with the air resistance. In order to take into account the air resistance force, you need to make Superman-b start off with a greater speed. However, if you start off with a greater speed, there is an even greater air resistance force. Once you start going up to super-fast speeds, the model for air resistance isn't really valid anymore.

Can Superman punch someone into space? Maybe, but it's not a simple punch. It's not even simple for Superman.

But what about the punch? Say that Superman hits Superman-b really hard. So hard that Superman-b has a speed of 10^5 m/s. What would happen? Let's say that the punch is right on the chin—an upper cut. Here is a diagram of Superman-b during that hit:

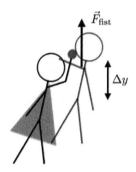

Here, Superman-b goes from a speed of zero to a speed of 10^5 m/s over a distance of Δy. What kind of force from Superman would this take? I will ignore gravity (its effect will be small in this case) and use the work-energy principle. If Superman-b is my object, then only Superman will do work. To calculate the work, I just multiply the force of the fist times the distance traveled. What does this work do? It changes the kinetic energy of Superman-b (it also changes the gravitational potential energy, but that term is very small compared to the change in kinetic energy).

Since I know the final speed and the distance over which he is punched, I can find the average force from this super punch. The only number I have not estimated is the distance over which the punch is exerted on Superman-b. I think 0.75 meters would be a generous estimation. With that, I get an average force of 4.67×10^{11} newtons.

Suppose that Superman's fist makes contact with a surface area of 70 cm² (I measured the front of my fist as an estimate—of course I made Superman's larger). What kind of pressure would this punch produce on Superman-b's skin?

$$P = \frac{F}{A}$$
$$P = \frac{4.67 \times 10^{11} \text{ N}}{0.007 \text{ m}^2} = 6.67 \times 10^{13} \text{ Pa} = 9.67 \times 10^{9} \text{ psi}$$

That's a high pressure. A typical scuba tank has 3,000 psi inside it and the steel tanks have a wall thickness of 1/4 inch. What am I trying to say? I am thinking that if Superman could hit Superman-b this hard, I think his fist would just push right through his head. That's gross, I know.

What about the pressure between Superman's feet and the ground? The force of Superman pushing on the ground would be somewhere around the same magnitude as the force he pushes on Superman-b. Of course, the contact area of his feet is probably higher than his fist, but the pressure would still be huge. I'm sure he would get pushed into the ground by his own punch.

What about the effect on Superman-b? If Superman-b has a mass of 70 kg, then I can get a value for his average acceleration during the punch. This would just be the force divided by the mass (again, the gravitational force is small in comparison). His average acceleration would be 6.67 x 10⁹ m/s².

What if I pretend like Superman-b is made of two parts: his head with a mass of 7 kg and the rest of his body with a mass of 63 kg. Superman pushes just on the head of Superman-b. Then why does the rest of his body also accelerate? Well, of course the head is connected to the body. This means that Superman-b's head pulls up on the body through the neck. In order for the body to have the same acceleration as the head, it would have to have a force of 4.2 x 10¹¹ newtons.

A *Nimitz*-class aircraft carrier has a mass of about 9 x 10⁷ kg. In order to produce the same force on Superman's neck, you could hang him upside down and then have 4,500 aircraft carriers hang from his head. I don't know about you, but I think his head would come off (also there aren't 4,500 aircraft carriers in the whole world).

So, let's be clear: Superman could not punch someone into space. If he did hit someone extremely hard, that target would probably lose their head (or maybe something worse). Superman would push himself into the ground during the punch as well.

Maybe Superman should just use his super breath to blow someone away.

CHAPTER 3: REAL STUFF

HOW MUCH ICE DO YOU NEED TO COOL YOUR BEER?

If you're going to spend some time outside on a hot day, it's nice to have a cold drink. What kind of drinks? Maybe it could be soda, maybe it could be beer. Either way, the best way to get your drinks cold is to put them in a cooler with some ice.

Here is the question:
How much ice do you actually need to get your drinks cold?

Let me start with some assumptions.

- Suppose you get *n* drinks and these start at room temperature. Let me say room temperature is 22 °C (about 72 °F).

- You start with ice and drinks. The ice is just at 0 °C.

- The cans are filled with water. I am actually surprised that canned water isn't more popular. Think about it. Okay, but assume it's filled with water, so I can then use the specific heat capacity of water.

- How much water? Well, the standard size is 12 fluid ounces. This would be 355 ml or 355 grams of water.

- The can is aluminum and about 15 grams.

- The cooler has no mass. Yes, it is one of those massless coolers that you can get from the store. Also, the amount of energy transfer while the drinks are cooling is small.

So, how will this work? What physics principles are involved? Let me start by saying that things have thermal energy. The hotter they are and the bigger they are, the more thermal energy they have. What I want is to transfer thermal energy from the drinks to the ice. This is one of the cool things about temperature: when you leave stuff in contact for a while, they reach the same temperature. Be careful, don't confuse temperature with thermal energy. If you put some pizza on aluminum foil and heat it up in the oven, both the foil and the pizza will eventually reach the same temperature. Even though the pizza has the same temperature as the foil, you can easily burn yourself on the pizza since it has much more thermal energy. Also, the pizza is much tastier than the foil.

If you put a drink in zero degree Celsius water, the first thing it will do is change phase from a solid to a liquid. This phase transition requires energy input into the ice. After that, the water (that was ice) will increase in temperature while the drinks decrease in temperature. At the final point, there will be drinks and water. This probably isn't what you want, but it would do the job.

How much energy is associated with a change in temperature? It turns out that the change in thermal energy for an object depends on the change in temperature, the mass, and the specific heat capacity.

$$\Delta E_{\text{thermal}} = mC\Delta T$$

Here, m is the mass of the thing, ΔT is the change in temperature, and C is the specific heat capacity of the thing. Also, different things have different specific heat capacities. This is why a hot, foam-based coffee cup does not burn you but the coffee inside (which is about the same temperature) does.

If something is changing phase, like from a solid to a liquid, then this also takes energy. The amount of energy needed for a phase change depends on the mass and the latent heat of fusion.

Now for an estimation. Let's say I have one can of soda or beer. How much ice do I need to cool that off? Well, how cool do you want it? If you can't decide, that is okay. I will make you a nice plot of final temperature of drink versus the amount of starting ice. Remember, I am assuming the drink (and the aluminum can) start at 22 °C.

The key here is that the change in energy of the ice (turning to water) plus the change in energy of the drink must be zero.

The problem is with the change in energy of the ice. If you assume all the ice melts and all of this energy comes from the decrease in thermal energy from the drinks, the drink could end up colder than the starting temperature of the ice. And while this is okay in terms of conservation of energy, it just doesn't happen

This is the key. In situations like this, the objects change temperature until they all end at the same temperature.

So, by using the above ideas, I can make a plot of the final temperature of your one drink as a function of the starting mass of ice (at zero degrees Celsius).

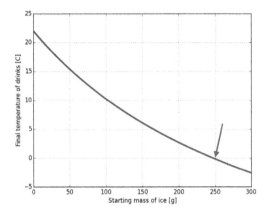

The arrow indicates the lowest temperature you could get the drink—it wouldn't get colder than the starting temperature of the ice. This says that with 100 grams of ice, you could end up with a mixture of zero degrees Celsius of water with your drink. Expanding this to a six-pack (of drinks, not my stomach) you would need 600 grams of ice.

Let me make this a little bit more realistic. The above calculation assumes all the thermal energy from the drink goes into the ice water. In reality, some other thermal energy will go into the ice (from the cooler and from the outside of the cooler). Suppose the ice gets 60 percent of the energy from things other than the drinks. In that case I would need 250 grams of ice per drink, or 1.5 kg for a six-pack and 3 kg for a twelve-pack of drinks.

What if I think of this in a different way? Instead of determining the amount of ice, suppose I purchase a 10-pound (4.5 kg) bag of ice? How many drinks would this cool down? Using the same calculations above, I get eighteen drinks.

So, what is the answer? I think I would recommend one 10-pound bag for every twelve drinks. This way, not all the ice will melt and you can keep your drinks cool for a longer period of time.

CAN A BUILDING BE A SUN-DEATH RAY?

This was a popular news story some time ago. Vdara Hotel in Las Vegas was reported to be a solar death ray. Basically, a curved, shiny building makes a solar hot spot from reflected sunlight. It would be a little like using a magnifying glass to concentrate enough sunlight to burn ants, except maybe you could incinerate cars and people. How would this work? Basically, the building acts like a two dimensional curved mirror. When light from a source really far away (like the Sun) hits a curved mirror, all of the light reflects to the same point. We call this the focal point.

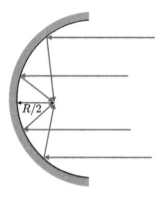

If parallel light rays come in (like from a faraway source) then they will focus at a point that is a distance half the radius from the center. But the building isn't a two-dimensional structure. It would be like a big, curved cylinder, not a line. If I want to model this building, I have to break it up somehow.

What if I look at just a single vertically standing rectangular mirror? I can then put a whole bunch of vertical mirrors along a curved path on the ground and boom, I have a building.

For one single rectangular mirror, the sunlight would cast a rectangular reflection. The area of this light on the ground would also be rectangular if the plane of the mirror was facing the sun. The size of this illuminated area on the ground would depend on the size of the mirror and the angle of the Sun above the horizon. When the Sun is lower in the sky, the reflected area would be larger. When the Sun is very high, the illuminated area would be small.

Why is the area of the reflected light important anyway? Intensity, that is why. Intensity is essentially the energy of light per second per area. Sunlight hitting the surface of the Earth is about 1,000 watts per square meter.

If the mirror is perfectly flat, all of the energy of the sunlight hitting the rectangular mirror would be evenly distributed over the reflected area. So, a small reflected area can lead to much higher light intensities than just from the Sun.

Now remember, that is the reflected light from just one rectangular mirror. The key here is that this building is like many of these mirrors arranged in such a way that a portion of the reflected light overlaps.

In order to estimate the intensity of light at this overlap point, I need some details about the building. Let me estimate that the building has a length of about 300 feet. I could break this length into thirty mirrors, each ten feet wide. If I also assume 70 percent of the sunlight is reflected by the building and the area of the "hot spot" is about ten by fifteen feet, then the intensity of reflected sunlight could be around 17,000 watts per square meter.

What would 17,000 watts of power per square meter do? Well, it would easily melt plastic. What about those solar hotdog cookers? You remember those? Basically, it is a cardboard box with reflective coverings on the inside. One side of the box has a clear plastic cover. Sunlight comes in and heats up the inside. For something like this, it works just like the greenhouse effect. The visible light from the Sun can pass through the plastic cover. However, once it heats up a hotdog, the hot hotdog gives off infrared radiation. This infrared light does not go through the plastic.

What kind of power per square meter does something like this solar oven produce? Well, if you assume that there is no focusing from the sunlight, it would just be the standard 1,000 watts per square meter. If you assume that all of this sunlight gets focused on something like a hotdog, the power per area would depend on the size of the oven and the size of the hotdog. Let's pretend the oven has a side that is 30 cm by 30 cm and the hotdog is 1 cm by 10 cm. This would give an area ratio of ninety to one. So, if all of the energy went to the hotdog, it would have 90,000 watts per square meter. I would guess the actual value is somewhere around 20,000 watts per square meter, which puts it right in the range of the sunlight from the building.

Can a building be a death ray? Yes, it can. I guess designers don't always take into account interactions with sunlight when they build these structures.

HOW CAN A BROOM BALANCE ON ITS BRUSHES?

If you've never seen it done, it's possible to balance a broom on its brushes. It's a cool party trick, taking your hand off it and letting it stand perpendicular to the ground, but the big problem is what people say.

> *"Hey, today is special because the planets are aligned and you can balance a broom!"*

Well, today may indeed be special (it could be your birthday or something), but the position of the planets has no effect on anything.

Let me start with gravity. Not your dad's "mass times g" gravity. No, the *real* gravity. Newton's gravity (unless your dad was Newton, then these two are the same thing). Gravity is an interaction between objects with the property *mass*. It is not just an interaction between things and the Earth. That just happens to be the thing with the most obvious interaction. Suppose I have two objects, mass one and mass two, that are separated by a distance *r* (as measured from the centers of the objects). The magnitude of the gravitational force between these two would be:

$$F_G = G \frac{M_1 m_2}{r^2}$$

The two m's are the masses of the objects and G is the gravitational constant with a value of 6.67 x 10^{-11} N*m²/kg². Sorry, I had to put down that constant. The important thing is that G is a pretty small number.

What about the broom? I will estimate the broom's mass at one kilogram. This will be important later. What objects could be interacting with this broom? Well, obviously the Earth. The Earth has a mass of about 6 x 10^{24} kg and the broom is about 6,000 kilometers from the center (the radius of the Earth). Now that I know all the values to put into the expression for gravity, I find a gravitational force of 9.8 newtons (I cheated and used rounded values for the mass and radius of the Earth, but it does work). You know why that looks the same as your "mass times g" formula? Because it is. Where do you think g = 9.8 N/kg comes from?

Now, how about a couple of planets? Right now, Venus is fairly bright in the night sky. But how far away is it? This is a perfect job for the Internet. In this case, I would recommend WolframAlpha[2]. It gives both the mass of Venus and the distance to Venus. Using these two values instead of the mass and distance to the center of the Earth, I get a gravitational force of 2×10^{-8} newtons. This force is indeed tiny compared to the gravitational force from the Earth. Why? Although the mass of Venus is very similar to the mass of the Earth, it is much farther away.

What about a different planet? Maybe one with a little more mass like Jupiter? Jupiter has a mass about a thousand times greater than the mass of Venus. Of course it is also farther away. Looking up exact values for the mass and distance, I get a gravitational force of 2×10^{-7} newtons. This is still tiny (in case you can't tell).

Let's do one more object. What is the gravitational force between you and the broom? Let's say you have a mass of 65 kg with a distance of maybe 0.3 meters from the center of yourself to the center of the broom. This would create a gravitational force of 4.8×10^{-8} newtons. Yes, this is also tiny. But look, the gravitational force from you is greater than the gravitational force from Venus. So here is your answer. How could the alignment of the planets matter when there are people around the broom that could matter almost as much (or maybe more)?

Then how do you balance the broom if it isn't due to the gravitational forces of the planets? It isn't difficult. There are two important things to consider. First, the center of mass of a broom is quite low. Much closer to the ground than many people would estimate. Since the brush part is at the bottom and bigger than the handle, the center of mass is low.

What does the center of mass have to do with balancing a broom? If the center of mass for the broom is not directly over some part of the support for the broom, it will fall over. In this case, the support area of the broom is covered by the brushes. There is another thing that is probably important: the brushes bend and act like a springy-type restoring force. This means that you don't exactly have to get the thing balanced before you let go. You just have to be close. Try it. Once you get the broom to work, you can try another version of this exact same thing: balancing an egg on its end. It has nothing to do with the Moon or the Sun or the planets. It can be done if you are careful on any day of the year. The key is to play around with the egg. They usually have these small bumps on the shell that can be used to make the egg stay up right. It looks cool, but it isn't as difficult as it sounds.

---

2   http://www.wolframalpha.com

Balancing a broom is cool and a fun trick, but it's not caused by the alignment of the planets. You don't have to believe everything people tell you. Sometimes, they don't even know the truth.

COULD ADAM SAVAGE HAVE ACCIDENTALLY KILLED HIMSELF IN THE SWIMMING CAR MYTH?

In one episode of *MythBusters*, Adam and Jamie revisited the ways you could escape from a car that was sinking in a river or lake. I have to say, this episode was quite exciting. They put Adam in a car (with a safety diver) and dumped it in a lake. There were cables to prevent the car from going deeper than fifteen feet, but this was still a rather scary myth. I consider myself to be fairly comfortable in underwater situations, but I was nervous for Adam in this case. It just looked scary.

In the first shot they showed off Adam escaping the car, everything looked fine. However, it was then revealed that he cheated. He had to use some of the air from the emergency diver. This is the part that also worries me. One of the most important rules when scuba diving is don't hold your breath while ascending. I am not saying Adam held his breath, it just wasn't obvious whether he did or not. Just to be clear, if you crash your car into a lake, hold your breath, and swim up, that's fine (well, at least that part is fine). The problem is breathing from a scuba tank and *then* holding your breath while ascending.

Why is there a "no breath-holding" rule in scuba diving? Recall that pressure is the force per unit area (force divided by area). As you go deeper in water, the pressure of that fluid increases. Why does the pressure increase? Well, there are several ways to think of this. Suppose we think about this in terms of floating. A block of water in water should float, right? Here is a picture of water floating in water:

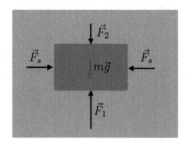

As you get deeper in water, the pressure increases. Assume that I have a balloon with water in it. Suppose I put this balloon in some water and pull it under. It will look the same as it does above water, with the same volume. This is because water doesn't compress that much. However, if I fill it with air, the balloon will compress. The lower I get, the higher the pressure and the lower the volume of the balloon. The size of the air inside shrinks until the pressure inside is equal to the pressure outside.

As you take the balloon down deeper, the pressure increases and the volume decreases. Now imagine this balloon is your lungs. Really, they are quite similar. If I take a deep breath at the surface and go down to a depth of 5 meters, my lungs will decrease in volume (because there is a finite amount of air in them). This actually happens.

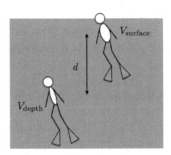

Now what if you do something different? What if you go down 5 meters and breathe from a scuba tank? One of the important things about a scuba regulator (the thing that attaches to the tank) is that it regulates the pressure of the air getting to the mouth. It provides air to the diver at about the same pressure as the water. Does this matter? Yes, it matters. The next time you go to a pool, try this. Take a two-foot-long pipe (PVC or something similar is fine) and go all the way underwater with one end of the pipe in your mouth and the other out of the water. Try to breathe. It is no simple task. Why can't you breathe through the pipe? Here is a picture:

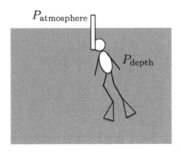

When you inhale, you want your lungs to expand. The problem is that since the pressure outside is greater than the pressure inside your lungs, your muscles have to really push. If your lungs don't expand, you can't bring in more air. It's like someone is sitting on your chest. Bring back the scuba regulator and it's quite easy to breathe since the pressure outside and inside your lungs are about the same no matter the depth. This is why I tell new scuba divers that breathing from a regulator is way easier than breathing through a snorkel.

I still haven't answered the question, have I? Why can't you hold your breath while scuba diving? How about we go back to Adam. Suppose he is 5 meters underwater in the inverted car. He is stuck, so he takes a couple of breaths from a scuba regulator. The pressure of the air in his lungs is the same as the pressure of the water at the 5 meter depth. Now what happens if he ascends while holding his breath? The opposite of the free diver going down. Instead of his lungs getting smaller, they would get larger—if only they could. However, they probably can't get any larger, especially if he took a full breath. This means that the lungs themselves have to exert extra pressure on the air and this only goes so far.

An ascending, breath-holding diver can have one of two very bad things happen. The first is an air embolism. Basically (and I am not a medical doctor here, so there's that) air from the lungs gets pushed into the bloodstream. Air bubbles in the blood are bad. These bubbles can cause all sorts of bad problems so let's just leave it at that. The second problem is called pulmonary barotrauma, where your lungs break or tear. Again, this is not a good thing.

Can divers hold their breath while using scuba gear? Sure, as long as they don't ascend while doing this. The consequence of holding a breath and ascending is high enough that divers are usually told "just never hold your breath." If you need to ascend while not breathing, breathe out at least. This will allow the expanding air in your lungs to escape. Actually, the common recommendation is to do something similar to a slight humming sound. This will allow the air to escape. While it can be an unnatural thing to do, you don't want to die, so you should keep the air in your lungs.

I am sure Adam's safety diver explained all of this to him in advance, but it still scares me.

HOW DIFFICULT IS A HUMAN-POWERED HELICOPTER?

Is it possible to build a helicopter that can hover while only being powered by a human? Well, yes it is possible. The Gamera II[3] project at the University of Maryland did just that. Their helicopter managed to hover for about fifty seconds. The obvious question is how hard would this be to accomplish?

No one ever said helicopters were easy. Have you seen the controls in those things? They are crazy complicated. But just because something is complicated doesn't mean we can't try to make a simpler model for it. In this case, I want to look at the power needed to make a helicopter hover.

At the most basic level, a helicopter hovers because it "throws" air down. It's almost the same way a rocket works, except it doesn't carry the air with it. If you throw a ball downwards, you have to push on the ball. This means the ball pushes back on you. If you could push hard enough on the ball, it could push back hard enough to keep you off the ground. Of course, the problem with the ball is that if you push hard enough it will get out of your hand very fast. If you have ever fired a gun and felt the recoil, that's the opposite and equal reaction of throwing the tiny bullet from the gun incredibly fast.

The helicopter solves this problem by throwing many "balls" down. Except in this case, the balls are air.

Let's make a model for this helicopter throwing down air. I will assume that the rotors on the helicopter push down a cylinder-shaped volume of air with some final speed (v). I can simply pick the length of this column of air and then determine the mass of the air based on the density (around 1.2 kilograms per cubic meter). If this air starts at rest, the force needed to get it up to speed would depend on the time it takes to speed it up.

Since the length of this air column depends on the speed and time, I can get an expression for the force from the air that only depends on the "thrust" speed.

$$F_{\text{air}} = \frac{\rho A v^2}{2}$$

Here ρ represents the density of air. A is the area that the rotors cover (and thus the area of air that is pushed down) and v is the speed the air is pushed down (which I am calling the thrust speed). Why the "2" on the bottom? This is

---

3   http://www.agrc.umd.edu/gamera/gamera2/index.html

because I assumed the air starts at rest and ends at a speed of *v*. The average speed of the column of air would be *v*/2.

If you want to check, this model for the force of air has the correct units for a force (newtons). Also, if you increase the size of the rotor or the speed the air is pushed, the resulting force will be greater. Both of these make sense. It would be crazy if the force was smaller with a higher speed, wouldn't it?

This force needs to be enough to support the weight of the helicopter. It also shows an important result. If you know the mass of a helicopter and the size of the rotors, you can calculate the thrust speed.

But what about the power required? If you recall, power is the amount of work done over the time it takes to do that work. The work will be in increasing the kinetic energy of this column of air from rest to the thrust speed. How long this takes also depends on the speed of the air. Putting this together, I get the following expression for the power needed to hover.

$$P = \frac{\rho A v^3}{4}$$

Remember, this is just a model for the power needed to hover. It doesn't take into account things like ground effect, forward motion, or other crazy stuff. But how can you tell if a model is reasonable or not? How about I use this model for some real helicopters? If you look at Wikipedia, you can find a lot of useful data. In this case, I can find the mass, rotor size, and engine power for a wide variety of helicopters.

Since the thrust speed can be calculated from the rotor size and the mass of the aircraft, I can use this to get an expression for required hovering power that only depends on the mass and the rotor area. Here is a plot of my calculated hovering power versus actual listed engine power:

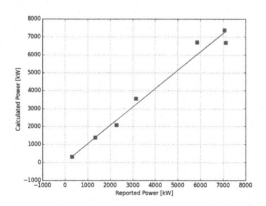

Looks pretty linear, doesn't it? This implies that the model I created for the power needed to hover isn't completely crazy. And there is one other cool thing about this real data from helicopters. If you plotted the calculated thrust speed versus the helicopter mass you will find almost no relationship. In fact, all of these helicopters have about the same thrust speed of around 28 m/s. That means the bigger the helicopter, the longer the rotors have to be, because the speed the air is pushed down is usually similar.

So how do we use this to build a human-powered helicopter? First, let's examine some real data from the Gamera II project. The human-powered helicopter (or humacopter as I like to call it) has a mass of 32 kg and consists of four rotors. Each rotor has a radius of 6.5 meters. For the mass of the pilot, I think 60 kg would be a good guess. I encourage you to go see footage of the machine, but I will sum up the important points. First, four forty-foot rotors means this thing takes up a huge amount of space. It looks like a giant pond skimmer made out of farm equipment. But it must be astonishingly light. Also, by flying for fifty seconds, we mean it hovered just off the floor indoors, under ideal conditions, and for less than a minute. This is not something James Bond or Batman would find useful.

The large rotor area is important. With a large area, the thrust speed can be much lower. There are two things that affect the power: rotor size and thrust speed. The power is directly proportional to the area of the rotors, but proportional to the thrust speed raised to the third power. This means that if you double the thrust speed, you increase the power by a factor of eight. Doubling the area only doubles the power. So, it makes more sense to have larger rotors with a smaller thrust speed.

If I use these values to calculate the required thrust speed, I get 1.68 m/s (3.8 mph). The power to go along with this thrust speed is 755 watts, just over one horse power. This is a pretty high value, but not incredibly high. According to Wikipedia[4], elite cyclists can obtain a power output of up to 2,000 watts for short periods of time. 755 watts would be tough, but possible. Plus, the pilot of this humacopter is using arms on a crank as well as his legs.

I can't resist pointing out that the humacopter also looks like the S.H.I.E.L.D. helicarrier, with its four large rotors. What is a helicarrier? In *The Avengers*, there is a flying aircraft carrier called the helicarrier. It appears to fly using four very large rotors. Could this design work? Could this aircraft even hover? Let me start with some assumptions.

---

4   http://en.wikipedia.org/wiki/Human-powered_transport

- I will use the helicarrier as seen in *The Avengers*. There are other variations of it in the comics.

- There are no special aerodynamic effects, such as ground effects, to help the helicarrier hover. For a helicopter close to the ground, it doesn't require as much power to fly. This is because the air from the rotors interacts with the ground which then interacts again with the helicopter. This is called the ground effect.

- The helicarrier in the movie is about the size and mass of a real aircraft carrier.

- The helicarrier stays in the air just from the rotors. It doesn't float like a lighter-than-air aircraft. I think this assumption goes along with the movie since they show it sitting in water floating like a normal aircraft carrier.

If the helicarrier has the same length and mass as a *Nimitz*-class carrier, it would have a mass of about 108 kilograms and a length of 333 meters. This would put the total rotor area at 4,000 square meters.

Using the same model as before, I can put these values in to obtain an estimate of the required thrust speed. With this, the air coming out of the rotors would have to be 640 m/s (1,400 mph). Just to be clear, this is faster than the speed of sound, although not as fast as the gas from the rockets on the Space Shuttle's Solid Rocket Boosters. Gas from rockets like the Solid Rocket Boosters typically has a speed between 5,000 to 10,000 mph. You can see the problem with having rotors that are too small. In order to fly you would need high thrust speed. Remember, the real helicopters had a thrust speed under 30 m/s.

What about the power? This is another problem. With a high thrust speed will come a very high power requirement. In this case, the power needed to hover would be 3.17×10^{11} watts (4.26×10^8 horsepower). That's a lot of horses. Just for comparison, the *Nimitz*-class carriers have a listed propulsion of 1.94×10^8 watts. I assume this is the maximum power, so it wouldn't be enough to lift the helicarrier. Obviously, the S.H.I.E.L.D. helicarrier has a better power source. I would guess it would have to be at least around 2×10^9 watts in order to operate. You don't want to use your maximum power just to sit still.

What if I want to fix the S.H.I.E.L.D. helicarrier? What if the rotors produced a thrust speed of 50 m/s (which is still quite a bit larger than real helicopters)? In this case, the rotor area would have to be 650,000 square meters. The rotors would have to go from a radius of 18 meters (as seen in the movie) to a radius of over 220 meters. Yes, that would look funny.

What if we went the other way? With the rotors at the size shown in the movie, how much could we expect them to lift? Using the same calculation but solving for the weight gives an aircraft mass of about 600 thousand kilograms or 1.3 million pounds. This is about the mass of a hundred-foot-long tugboat. A tugboat-sized helicarrier wouldn't be that cool.

In the end, flying a human helicopter looks difficult, but is possible. Well, of course it's possible since it's actually been done. If you want to make something giant fly with helicopter blades, it better have a massive power plant or very large rotors.

HOW MUCH WATER WOULD IT TAKE TO MOVE A CAR?

Sometimes bad things happen. A large flood could occur, and it would certainly suck. During floods, you often see cars that were moved by rushing waters. But how much water would it take to actually move a car?

First, why would a car move? If the car is at rest, it would have to go from a zero velocity to a non-zero velocity. This means there must be a net force that is non-zero to get the car moving. A big caution here: you need a net force to get it to move, but not to keep it moving. If it is moving, a zero net force will be fine.

Suppose there is a car in some running water. There are five significant forces to consider. First, there is the gravitational force. Basically, this has a value of the mass of the car times the gravitational field (9.8 newtons/kilogram). The gravitational force is pretty straightforward. The only way to change the gravitational force is to change the Earth. Let's just assume that the Earth stays as it is, even though there is a flood.

The next force to consider is the force the ground pushes up on the car. This is called the normal force since it is perpendicular to the ground. The only exciting thing about the normal force is that it will push up with as much force as possible to keep the car from falling through the ground. For a typical car at rest on a flat road, the normal force would have the same value as the gravitational force to give a net force of zero.

What about the water? Would the water exert forces on the car? Let's say there is water of some depth, h, moving at some speed, v. This water will push on the car in a very similar way to air pushing on a moving car. In fact, I will use the same model that I have used for air resistance. This model says the force will be equal to the product of the following terms:

- The density of the water
  (which is pretty constant at around 1000 kg/m³)

- The size of the surface area that the water pushes on
  (not all of the car would have to be underwater)

- The shape of the car

- The square of the speed of the moving water

So, faster-moving water would push hard on the car. This seems obvious. However, deeper water will also push more on a car since more water will be colliding with the car.

Of course, just because there is water pushing on the car doesn't mean it will start moving. Why not? Friction can prevent the car from sliding. There is a frictional force between the tires and the road. The most basic model of friction says the maximum frictional force depends on two things. It depends on the types of material interacting, and in this case that would be wet rubber on asphalt or cement. The other factor is the force with which the two surfaces are pushed together. This would be the same as the force the ground pushes up on the car. Notice that the maximum frictional force does not depend on the size of the contact area between the tires and the road. This is just an approximation. Of course if you have large tires it might make a difference, but these are just going to be normal tires.

There is one other very important force: the buoyancy force. As the water level rises, the water will push up on the car. What will this do? It will mainly reduce the force the ground pushes on the car. With a smaller normal force, the maximum friction force will also be smaller. But how do you calculate the buoyancy force? The simplest method is to look at the volume of water the car displaces. The buoyancy force will be equal to the gravitational force of this displaced water. The car doesn't have to be fully submerged to have a buoyancy force. If you have been in a pool, you will probably realize this. Even in the shallow end of a pool, it is easier to lift up another person. This is because of the buoyancy force.

Now it is time for some numbers. Let me pick a car. I have a Toyota Sienna minivan, so I will choose that. If it's just sitting there on a flat road, what is the maximum frictional force from the tires? Well, first I need the mass of the car. The listed mass of this vehicle is around 2,000 kilograms. Now, what about the tires? Let me use a coefficient of friction with a value of

0.4 assuming the road is wet, but not flooded. This means that the maximum frictional force on this car would be almost 8,000 newtons (1,700 pounds). You would have to push this hard in order to get the car to slide on a wet road.

If the water level was up to 0.5 meters deep, the car would probably have a significant buoyancy force on it. If I had to guess (and I do), I would say the main, floaty part of the car is about 0.3 meters above the ground. This means for half-meter-deep water, the car would have to be 20 cm underwater. Of course, this also assumes the car doesn't leak (it will), but if it is slow enough it won't matter. If the car is 5 meters long and 2 meters wide, this would make it displace about 2 cubic meters of water. Since water has a density of 1,000 kg/m³, the mass of this water would be 2,000 kilograms, producing a buoyancy force of almost 20,000 newtons. This would be enough to just make the car float, so I guess it could easily move.

What about some lower level of water? What if the water was just 0.4 meters deep? How fast would the water have to be moving in order to get the car to move? All I have to do is to set the "water resistance" force equal to the maximum frictional force. Just to be clear, both of these will depend on the depth of the water. The only value I haven't already declared is the drag coefficient. I chose to use a value of 1.0, which is close to the aerodynamic drag coefficient of a cube. Yes, I know that drag from water and drag from air are different things. However, that won't stop me from using this model to get a rough estimate of this force.

Now that I have all the values, I get a water speed of 6.2 m/s or 13 mph. That's some pretty fast water. But what if I just increase the water level by 5 cm? Putting in these new values, I get a minimum water speed of 3.6 m/s (8 mph). That's still pretty fast, fast enough that you probably couldn't stand up in this kind of water.

Could this really happen? Of course. I suspect you can get this kind of high-speed water when something dramatic happens such as a break in the levee or water bursting out of an above-ground pool. We also know this happens because you have seen it on the news and on funny YouTube videos.

CHAPTER 4: STAR WARS

THE POWER SOURCE FOR A LIGHTSABER

This one has been on my mind for quite some time.

> *What kind of power source would you need to run a lightsaber?*

I was actually worried about this post when I saw the Discovery Channel show *Sci Fi Science*. In one particular episode, Michio Kaku, theoretical physicist and futurist, talks about how you could actually build a lightsaber. The episode was a little silly but the science wasn't too bad. In the end, Michio decides to build a type of handheld plasma torch. In doing this, he estimated that the lightsaber would need a power source on the order of megawatts.

He didn't do what I was considering examining. I am thinking about the scene from the beginning of *Star Wars Episode I: The Phantom Menace* where Qui-Gon tries to cut through a door using his lightsaber. I can use this cutting action to estimate how energy would need to be stored in the lightsaber since the metal is essentially melted. How much energy would that take?

Let me go ahead and acknowledge that lightsabers aren't real. Oh, maybe they are real, but they use some special Force crystals. Of course, this won't stop me from making an estimate anyway.

Before I get to the estimation, how about some background ideas? First, let's talk about blackbody radiation.

A blackbody is an object that gives off light due to its temperature, not because light is reflecting off of it. When things are hot (and even when they are not), blackbodies give off electromagnetic radiation. This is true for dense solid objects, not for low density gases. These objects give off a wide range of "colors" of light with the wavelength of the peak of the distribution related to the temperature of the object.

When an object gets warmer, two things happen. First, more light is given off by the object. Second, the color of the highest intensity light shifts more towards the blue part of the spectrum.

There are two excellent everyday examples of blackbodies: the Sun and an incandescent light bulb filament. Another example is the element on a hot stove. As the stove element heats up, it produces light mostly in the infrared region (so you can't see it). As the element gets hotter, this spectrum of light it produces shifts towards shorter wavelengths and starts to look red (but don't touch it!). If it gets even hotter it will start to look yellowish.

The point is that you can determine the temperature of a blackbody by the color of light it is giving off. Let me leave it at that (although it is possible to make it much more complicated). In the clip from *The Phantom Menace*, I will use the color of the hot door to determine its temperature.

Now, what about thermal energy? How much energy does it take to increase the temperature of a material? Well, this depends on the change in temperature, the mass of the object, and the specific heat of the material.

We often say the change in thermal energy of an object is the same as the heat going into the object (heat is represented by the letter Q usually). When discussing this thermal energy stuff, I like to talk about a pizza in an oven on some aluminum foil. Suppose you put the pizza in the oven until the temperature is 350 °F. Can you touch the aluminum foil? Yes, but don't touch the pizza. The aluminum foil has a very low mass and thus does not absorb and store much thermal energy (so it doesn't hurt your hand). This is not true for the pizza.

There is another thing to consider with thermal energy. What if the material changes phases, as in, it goes from a solid to a liquid? This also takes an amount of energy that depends on the mass of the stuff and the type of material. The energy needed to take a material from a solid to a liquid is the latent heat of fusion multiplied by the mass of the object.

Now for some measurements and estimates from the scene in *Star Wars*. By looking at a frame from the movie, I can adjust an online blackbody simulator[5] until it shows the same total color. This gives a metal temperature of about 2,700 K for the cooler parts of the door and maybe 5,200 K for the part right near the lightsaber.

What about the mass? This is a bit more difficult. First, the whole door gets hot (you know, from thermal conduction) so let me just start with the hotter parts. Qui-Gon makes a line with his lightsaber that is about 2 meters long and the width of his lightsaber (about 7 cm). How deep is this cut? Just by looking at

---

5   http://phet.colorado.edu/en/simulation/blackbody-spectrum

the lightsaber poking through the other side of the door, I am going to guess it is about 20 cm thick. This gives a total volume of 0.028 cubic meters. For the rest of the stuff in the door, let me just say that the cooler (but still hot) part is twice the mass of the melted part.

If I knew what type of material the door was constructed from, I could get the density and thus the mass. In considering the material, I can imagine someone getting upset like this:

> *"Hey Dude! How can you calculate the energy of that stuff? The Trade Federation totally stole some secret transparent aluminum material and it has a really low density!"*

I agree, this could be something weird. However, I don't know how to esti-mate the specific heat or density of weird stuff. Let me make the assumption that this is something like a known material. So, what material is metal and melts around 5,000 K or at least has a melting point greater than 2,700 K (because that hot part could be hotter than the melting point)? If it was made of an element, probably the best fit would be something like tungsten or carbon. However, those do not seem likely. Titanium melts at 1,930 K. If it is an element, I would pick titanium because it is awesome. The Trade Federation only uses awesome materials.

Throwing all these numbers together and assuming the cut takes about nine seconds, I get a minimum power output of the lightsaber at around 28,000 watts.

Now, where can I get 28 kilowatts? At least it isn't 1.21 gigawatts—that would require a lightning bolt or a Mr. Fusion

The power needed to cut that door tells me something (but not everything) about the energy source for the lightsaber. I want to estimate the energy den-sity of the energy source. To do that, I will estimate how long the lightsaber will run without recharging. This is a tough one. Maybe they run forever but I will not assume that because it wouldn't be as much fun. So how long would it have to run for it to be useful? I say at least two hours of continuous use. That seems reasonable, doesn't it? How much energy would that be? Because power (in watts) is the energy (in joules) divided by the time (in seconds), this would be a total energy of over 3 million joules.

In order to estimate the kind of power source in this lightsaber, I need to know how big it could be. Taking an estimate on the high end, I will say it is a cylinder with a radius of 3 cm and a length of 15 cm. I think that is certainly big enough. If this is the case, then the energy density of this power source would be 8 billion joules per cubic meter. Just for comparison, Wikipedia has a nice table of energy densities[6]. This puts the lightsaber source at somewhere between octanitrocubane explosive (no idea what that is) and beryllium plus oxygen (again no idea). As far as known Earth batteries go, it seems like the highest energy density is the fluoride ion with 2.8 MJ/L.

Perhaps you would like to run your lightsaber on plain old AA batteries? A nice AA battery has about 3 watt-hours of stored energy. If I want to run my 28,000 watt lightsaber for two hours, it would need 56,000 watt-hours of energy. So, how many AA batteries? How about over 18,000 AA batteries? I wonder if the lightsaber comes with batteries.

Okay, I get it. The Jedi have some secret, Force-related power source.

SHOULD BIGGER SPACESHIPS HAVE BIGGER THRUSTERS?

If you are a nerd and have been on the Internet, you've surely seen this awesome website showing the relative sizes of many different science fiction (and some real) spaceships (http://www.merzo.net/) created by Jeff Russell. Very cool stuff. I highly recommend you take a moment to look around the site.

Of course, when I see images like this I can't just say "cool." One of the things I always think is interesting is to consider objects of different size. Perhaps the general idea is that you can just scale up or down as you like. But it doesn't work this way. Let me start with my own spaceship. It is a sphere with a thruster on the back. It just holds one person.

---

6   http://en.wikipedia.org/wiki/Energy_density

Now, what happens if I want to make a bigger version? Let me go ahead and get some related points out of the way.

What does the thruster do? This is a great question. In real spacecraft, thrusters are used to change the momentum of the craft. You could think of these thrusters as exerting a force on the craft which would make it accelerate. What if you kept the thruster on for a long time? You would keep accelerating. Do you see the problem? Most of these ships in science fiction shows fly with their thrusters on at a constant speed. This is what would happen if you have some resistive force like air resistance.

The point is: I am going to assume the thruster accelerates the ship, which may not agree with the movie. Now I'm moving on.

How about some other assumptions? Of course, some science fiction fans might not agree with all of these, but that never stopped me before. Some of them may not actually be valid. However, they are close enough to be true that it will show my point.

- Force from the thruster is proportional to the area of the thruster. You could come up with all sorts of reasons why this would not be true, but I am still going with this assumption.

- The densities of big and small starcraft are about the same. Yes, maybe the walls of a ship are the same thickness, which would make bigger ships have a lower density, but they have additional internal walls as well.

- Small starcraft and large starcraft have thrusters to produce the same (or similar) acceleration.

- The following all mean the same thing: spaceship, starcraft, and spacecraft, but not starship, which we use only to denote the group that sang "We Built This City (On Rock and Roll)."

Now for some physics. Let me assume that if there is only the force from the thruster acting on the spacecraft, then the acceleration will be proportional to the force from the thruster. The acceleration will also depend on the mass of the spacecraft. For the same force, a larger mass will produce a lower value of acceleration. This is, in essence, Newton's second law of motion.

My make-believe, single-person ship has a radius R and a circular thruster with a diameter L. If thrust is proportional to the area of the thruster, then the force on the spaceship would be some constant times L squared.

What about the mass? Because this spaceship is spherical, the mass will be some constant (in this case that constant would be $\frac{4}{3}$ pi) times R cubed. Because the acceleration will be the force divided by the mass, this means it will be proportional to L squared over R cubed. I could say that the diameter of the thruster has to be proportional to $R^{3/2}$. The constants don't really matter.

Now, let me build an even bigger spaceship. It is going to be ten times the radius of the first one. It will have about the same density and be capable of the same acceleration. How big would its thruster be? If the other spaceship parameters are the same, and I increase R by a factor of ten then the diameter of the thrusters would have to increase by a factor of 31.6.

Just to be clear, making the spaceship ten times longer increases the volume (and thus the mass) by a factor of 1,000. In order to increase the thrust by the same amount, I have to make the thrusters more than ten times larger because thrust is proportional to area.

The result is that this bigger spaceship wouldn't look the same. Here is a diagram of the one-man spaceship and its longer counterpart side by side:

A bigger ship means even bigger thrusters.

Now, how about an example from *Star Wars*? There are two ships that make a great "case study" for thruster size: the Star Destroyer and the Super Star Destroyer. What makes these two ships great for a comparison? Well, they are in the same universe. They have the same shape. I can safely assume they have similar densities. Finally, if they are going to be in the same fleet, it seems reasonable they would have similar accelerations. You can see a side-by-side comparison on Jeff Russell's spaceship page.

Apparently there are some that debate the size of the Super Star Destroyer (for example: http://www.theforce.net/swtc/ssd.html). I will go with the dimensions from Jeff Russell's site. This gives a Star Destroyer a length of 1.6 kilometers and the Super Star Destroyer a length of 19.0 kilometers.

What about the thrusters? According to a page at TheForce.net, there are thirteen thrusters on the back of the Super Star Destroyer and three on the back of a plain Star Destroyer. By examining the diagrams, I also find that the diameter of the Star Destroyer thrusters is about 0.126 kilometers. Since I know the length and the thruster size, I can determine how large the Super Star Destroyer (SSD) thrusters have to be to have a similar acceleration.

Get ready for this. Taking into account the fact that the SSD has more thrusters, I find that each one has to be 2.48 kilometers in diameter. So what? That means instead of a bunch of cool-looking nozzles here and there, probably half of the total ship would need to be made of thrusters. Instead of an imposing death machine, it would look like one of those swamp fanboats.

HOW MUCH DOES R2-D2 WEIGH IF HE CAN FLY?

Let me start with a quiz question.

In *Star Wars Episode II: Attack of the Clones*, R2-D2 shows that he can fly. What is wrong with a flying R2-D2?

- (a) Nothing. This is the way George Lucas would have wanted it in the original *Star Wars* but he couldn't digitally render a flying R2 with a Commodore 64.
- (b) If R2 can fly, why didn't he do this in the original trilogy?
- (c) He wouldn't fly that way.
- (d) Droids shouldn't be allowed to fly.

Here you see this is really a trick answer for choice (a). Why is (a) a problematic answer? Because the Commodore 64 didn't come out until 1982, and *Star Wars* (it was just called *Star Wars* then) came out in 1977. So, choice (a) can't possibly be correct.

The correct answer is (c): R2-D2 wouldn't fly that way.

But how does he fly? If you watch the movie carefully, you will see that while R2 (all his close friends just call him R2 and not his full name of R2-D2) is flying, he points his thrusters partially backwards and partially down. Here is another robot that flies in a similar manner to R2-D2:

This is the exact same way R2 flies while he is moving at a constant speed. You might think: well, what is wrong with this? It looks perfectly fine, right? I think this is the real issue. R2 is shown to fly the way people think about forces and motion, so no one sees it as a real problem.

Ah, but now we are talking about forces and motion. Let me start with two experts in this field: Aristotle and Isaac Newton.

What does Aristotle say about forces and motion? For him, constant force means constant motion. Honestly, to most people this idea simply makes sense. It is easy to agree with Aristotle, even if he is speaking in Greek. Doesn't this idea of forces always seem to work? If I push a book on a table, the book moves. If I push harder, it moves faster. If I stop pushing the book, the book stops. The idea is simple.

According to Aristotle, R2-D2 flies as he should. If he wants to fly at a constant speed, he needs to angle his thrusters back a little. This way, part of the thrust pushes down to keep him up and part pushes back to move him forward.

Now, what about Newton? Well, it's not just Newton that came up with a better idea about force and motion. It's just that people call these ideas "Newton's Laws" of motion. Newton says that forces change the motion of an object. "Change" is the key word here. If you have a constant force on an object, it would constantly change its motion. This means that it could keep increasing in speed with a constant force.

Consider the following example. A bowling ball is sitting on a smooth bowling alley lane. If you give it a little push, it will start to move. It will keep rolling for a while, but it will eventually stop. This is because there is another small force on the bowling ball that most people tend to forget about: friction. So after the push, there is just one force acting on the ball to make it slow down. However, if you ran along behind it, you could keep giving it little pushes to keep it going the same speed, balancing the frictional force. If you used a long stick to keep applying a continuous level of force that was greater than the frictional force, it would go faster and faster.

Suppose there was a way to remove all forces from a moving object. For such an object, the speed would stay constant. I know this is hard to image because everything we see on the Earth has some type of frictional force on it.

Okay, so if you look at the way R2 flies, it seems to agree with Aristotle and probably 90 percent of the human population. But is there any way to make this flying R2-D2 agree with the Newtonian ideas of force and motion? In other words, in what situation would this type of flying agree with Newtonian physics?

Air resistance is our best bet. Suppose there is non-negligible air resistance force on the flying R2. In this case, he would have to have his thrusters angled back to balance the horizontal air resistance. We can actually use this to figure out how much R2 weighs. If we know what the air resistance and his speed are, we can figure out how much thrust he is pushing forward with. If we know how much thrust he is putting out, we can figure out how much he must weigh for that thrust to lift him. If I drew a diagram with the forces on R2, it would look like the following:

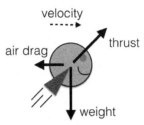

If I can model the magnitude of this air resistance force, then I can get an estimate of the amount of thrust and therefore the weight of R2.

First, some assumptions:

- I will assume Earth-like gravity. Why? Why not? If you look at the people in the movie they look like they are jumping around on the Earth (because they are).

- Earth-like air. This one is harder to justify. How do I know the air isn't super-dense or has a very low density? I don't. Since I don't have much choice here, I will assume the air on this planet has a similar density to that on the Earth.

- For the details of this situation, I will also use a flying speed of 2.3 meters per second. This isn't an assumption. By using video analysis, I can get a plot of position versus time for the flying R2. The slope of this line is the speed.

- The thrusters are angled back at about 43 degrees. This was also a measurement from the video.

- Finally, I will have to estimate some of the physical dimensions of R2-D2.

The next thing is to estimate the magnitude of the air resistance force. For objects at low speeds (like this), the air resistance is proportional to the square of the speed, the size and shape of the object, and the density of the air. With some estimated values for the size and shape, the R2 unit would have to have a mass of around 100 grams. This would put his density around the density of air itself. If the density were that low, he would float. Really, he wouldn't even need to point the thrusters down at all.

In the end, R2-D2 is probably not that light. He is just flying wrong.

HOW FAST ARE BLASTER BOLTS?

You have no idea how long I had been planning to look at the blasters in *Star Wars*. No. Idea. Finally, when the 35th anniversary of *Star Wars* came up in 2012, it motivated me to complete my study (which I hadn't actually started). Here is the deal: What are these blasters in *Star Wars*? How fast are the blaster bolts? Do the blasters from spacecraft travel at about the same speed as the hand-held blasters? Why do people still think they are lasers?

I am almost certain (*almost*) that nowhere in the *Star Wars* movies (even in *Episode I*) does a character refer to these as weapons as "laser guns." There isn't much to talk about here that hasn't been discussed a thousand times before, but in short, there are two points. First, if they were lasers, you wouldn't be able to see them from the side very well. You have a red laser pointer, right? You don't really see the beam unless it is going through something like chalk dust. I am sure some geek out there has a great explanation for why you could still see the beam (such as non-Earth-like atmospheres or maybe all the scenes actually take place underwater). However, it doesn't matter because you can't see lasers from the side.

The second point is that lasers travel at the speed of light. Yet clearly, you can see these beams have a speed that is much smaller than the speed of light. How much smaller? I don't know, but I am going to find out.

How do you find the speed of a blaster bolt? I don't think there is just one answer to this question. The answer depends on the particular scene showing the blaster bolt being fired. Well, let's get started then. The first scene in *Star Wars* shows Princess Leia's Blockade Runner trying to escape from an Imperial Star Destroyer. Of course, both spaceships are firing at each other.

How do you get the speed of these bolts? In general, if I know the distance the blaster bolt travels as well as the time it takes to travel this distance, I can find the average speed by dividing the distance by the time. You can get the time from a video by counting frames. If the movie is recorded onto a DVD, it would have a frame rate of 30 fps. This means that each frame is just 0.033 seconds apart. The first thing to consider is the size of the smaller spaceship.

Looking at this first scene in *Star Wars*, I need to get an estimate of distance. How far away is the Blockade Runner from the Star Destroyer that is chasing it? Is perspective a big problem? In this case, I am just going to make some wild estimates. Let me just go to the Starship Dimensions website[7]. Yes, this is the same awesome site that I used to look at the size of different spacecraft thrusters.

From the Starship Dimensions site, the Rebel Blockade Runner is listed as being 150 meters long. Because perspective and camera angle are a big problem here, I am just going to go with a distance of roughly ten Blockade lengths. In meters, maybe I could say 1,500 +/- 500 meters. That seems pretty far, but it's just an estimate.

Looking at each frame in the video with a traveling blaster bolt, I get a time of flight of 0.08 seconds. From the distance estimate, this puts the blaster bolt speed at 180 kilometers per second.

What about non-space shots? The very next scene shows Imperial Stormtroopers blasting into the Blockade Runner and showing the Rebels who's the boss. The analysis of this shot is a little different. The camera seems to be far enough away that maybe I could try a real video analysis. In video analysis, you can use some software[8] to mark the location of an object in each frame of a video. Once you have chosen a scale, the program will give you the x and y coordinates and time data for that object. In this case, I assumed the distance from the stormtrooper's belt to the top of his head was 0.71 meters (based on measurements of a full-standing stormtrooper with an assumed height of 1.78 meters). After getting the position and time data from the video, I get a bolt speed of 15 m/s. Maybe you can already see the problem here. The space bolts are way faster than the handheld bolts. Well, maybe this isn't a problem so much as further evidence that they are not lasers. Lasers would all travel at the same speed.

---

7   http://www.merzo.net/

8   I like the free and awesome video analysis software Tracker Video Analysis - https://www.cabrillo.edu/~dbrown/tracker/

Okay, those were just two examples. But there are *many, many* shots fired in *Star Wars* (I am leaving off the *"Episode IV: A New Hope"* part). Let me see if I can get more data from the rest of the scenes.

Before showing you the rest of the data, let me look at a special case: the Death Star. I am not sure if what's coming out of the Death Star is the same as "blaster bolts" or not, but I analyzed it anyway. If the Death Star has a diameter of 160 km, then I can get a rough estimate for the speed of the beam coming out as it destroys Alderaan (which is a peaceful planet without weapons). There are actually two different speeds of the Death Star's shot. First, something comes out of the massive circle on the Death Star. This stuff then combines together to make one giant beam.

After a quick analysis, I find that the first phase of the beam has a speed of 600 km/s and then once the beams combine, the resulting shot travels at about 1,000 km/s. Both of these values are obtained from the scene that shows just the Death Star.

Here is the odd part: in the next shot, the beam is shown traveling towards Alderaan (a peaceful planet). It takes about 0.2 seconds for this beam to reach the planet. If the speed of the beam is constant, this would make Alderaan only 196 km away from the Death Star. I'm not sure how big Alderaan is, but the International Space Station is about 300 km away from the surface of the Earth, so . . .

Now for the rest of the data from episodes *IV*, *V*, and *VI*. Why didn't I also include data from episodes *I*, *II*, and *III*? Well, many would argue that those aren't truly *Star Wars* movies, mostly because of the inclusion of Jar Jar Binks. However, there is another reason: these movies have many, many more blaster shots. There are too many for me to consider at this point. After going through the original trilogy, I would estimate that I have data on about 10 to 15 percent of all the blaster shots. There are many scenes in these movies that either don't show a full view of the blaster or don't have anything nearby to find a distance scale. Some of the blaster shots are moving either towards or away from the camera such that there would be significant perspective problems. So, in the end, I have data on ninety-one shots. Nineteen of these shots are "space" shots.

There is really only one way to show you the distribution of all the blaster bolt speeds. I am going to create a histogram, but a plain histogram wouldn't work too well. Why not? Because I am trying to represent speeds from about 10 m/s to about 1 million m/s. The plot would just look silly. Instead, let me make a histogram showing the natural log of the bolts' speeds, so it is not a linear scale.

Here is the plot showing the speed for ground-based, space-based, and Death Star bolts.

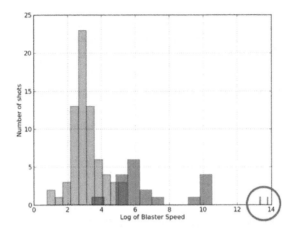

The circled area shows the data from the Death Star. I only have two values, but they are way out there in terms of speed. And yes, there is some overlap in speeds for the space and ground shots. Why? Well, there are a couple of far-away ground shots and a couple of close-up space shots (like when they show R2 in the X-wing fighter with shots whizzing past). But it still seems clear these ground and space shots are different.

One other thing I just noticed: When a spaceship fires a blaster, some will fire red blaster bolts and others fire green blaster bolts. The handheld blasters all fire red-colored bolts. I have no idea why there aren't green-colored bolts for the handheld weapons.

Why are the speeds so vastly different? Don't worry, I am not completely delusional. I know that *Star Wars* is just a movie. I know that Han Solo's blaster doesn't really shoot anything, except maybe blanks. Human beings have to actually draw these blaster bolts on the screen. Humans tend to draw things consistently on the frame, independent of the setting. If you instead did an analysis without appropriately scaling each scene, you would find all blaster bolts have about the same angular speed. That is, they appear to move at the same speed on the screen independent of the actual setting.

Fine, the artists in *Star Wars* are just humans making human mistakes. Yes, but is there any way to fix this? First, let me comment on the ground-based blaster shots. The average for them is just 34.9 m/s (78 mph). This is in the ballpark of a baseball pitch. Compare this to the speed of a Nerf gun's bullet at about 10 m/s.

This means two things:

- A Jedi deflecting blaster bolts with a lightsaber is about the same as a baseball player hitting a pitched ball.

- Playing with Nerf guns and plastic lightsabers in the backyard isn't too terribly different than the movie.

Actually, it wouldn't be too difficult for any normal person to dodge one of these blaster bolts, especially if it were fired from so far away. Maybe this explains why the stormtroopers suck so much at shooting. They don't suck, it's just that Han, Chewie, and Luke can easily dodge these bolts when far enough away. The stormtrooper, on the other hand, can't dodge. Why? Because those blasted helmets block their vision. You can't dodge what you can't see (well, except for Luke).

What about the discrepancy between space bolts and handheld bolts? I think this is mostly okay. They aren't the same weapons, are they? Because they aren't the same weapons, they don't have to have the same speeds. The only thing you would need to fix is to make the speeds of these space blaster bolts consistent. That means no more scenes showing shots flying past R2 at close range. The shots would just be too fast to see.

The other change would be to increase the speed of the blaster bolts from handheld weapons. If you wanted bullet-like speeds of around 500 m/s, what would change in the movie? Well, the first thing is there probably wouldn't be any two consecutive frames where you see the same bolt. That could be a simple change. If you see the gun firing, then don't show the bolt. If you want to show "whizzing by" bolts, just show one. That way, a blogger like me wouldn't have a very good method for determining the speed of the bolt. Problem solved.

One more thing. What is a blaster bolt anyway? It isn't a laser, right? My guess has always been that it is some sort of super-hot thing. Maybe it is a gas so hot that it has become a plasma. The problem with it being gas is air drag. If the bolt has a low mass, I suspect it wouldn't get too far (especially at those low speeds). Perhaps the gas is so hot that it ionizes the air in front of it. Or maybe it is some type of extremely small, hot bullet. Honestly, I am not sure.

You know what happens when you talk about *Star Wars*? Geekplosions. Let me go ahead and preemptively answer some of them. Please don't be offended by these fake comments, I am just trying to be funny.

- "Seriously? You wasted all this time analyzing something that was clearly not real?" **I am not sure that this is actually a question, but yes — it is true. I could say the same thing about you though Seriously? You just spent 8 hours playing a videogame? It isn't even real.**

- "You must have no life. Why don't you get out and do something?" **This is essentially the same as the first comment.**

- "You actually get paid to waste your time like this?" **I am not sure I get actually paid for this actual analysis. But I really do think this is a worthwhile post. It shows how to take data from something (even if that some thing is obviously fake) and analyze it**

- You said that the Death Star is 160 km in diameter. Actually, the original Death Star plans had it at 180 km diameter but it was changed to 160 km in order to finish on time to destroy the Rebel base on Yavin 4." **Okay.**

- "How can you hear the blaster bolts in space?" **If you use The Force, you can hear them.**

- "How much energy is in each blaster bolt?" **Another great question. You could estimate the volume of a bolt and assume it is gas at a certain temperature. You would have to guess at the density but that would be one way to get a value for the energy.**

- "I kind of thought you would do more with the uncertainty in the speeds of these bolts." **Me too.**

DID HAN SOLO SHOOT FIRST? COULD HE HAVE SHOT SECOND?

They make t-shirts that say "Han Shot First." In Geek/Nerd culture, this is a big deal. Let me give you the short version. In 1977, George Lucas released the first *Star Wars* movie. At one point, Han Solo is confronted by a bounty hunter named Greedo because Han owes money. Greedo pulls out a blaster and threatens to shoot Han. But wait! Han beats him to it and pulls out his own blaster under the table. Blam! Greedo is dead.

For most *Star Wars* fans, this says something about Han's character. It says he is a survivor. He will do what it takes to get through the day. It also shows that he has experience. He knows this isn't going to turn out favorably unless he takes action. So, action he takes. Does it make him a bad person? How about badass? Han isn't a Lawful Good, that is for sure. Poor Greedo. He never had a chance.

In 1997 George Lucas re-released *Star Wars Episode IV* with a few changes. Sure, there were more special effects, but the Han/Greedo scene was also changed. In the new version, Greedo shoots at Han but misses (and only from about 2 meters away). Of course Han has no option but to return fire, it's just self-defense, right?

Next, George Lucas re-released *Star Wars Episode IV* in 1997 with a few changes. Sure, there were more special effects, but the Han/Greedo scene was also changed. In the new version, Greedo shoots at Han but misses (and only from about 2 meters away). Of course Han has no option but to return fire. It's just self-defense, right?

Let's take a closer look at the sequence of events in the 1997 version. Using video analysis, I get the following plot for the motion of Han's and Greedo's blaster shots:

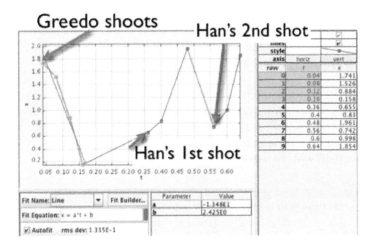

Greedo shoots — Han's 2nd shot — Han's 1st shot

The first thing you might notice is that Han shot twice. One shot isn't good enough for Han, he wanted to make sure that Greedo was a goner.

Using the above video analysis, I get the following timeline of events:

- At t = 0.04 seconds, Greedo shoots.

- At t = 0.36 seconds, Han shoots his first shot.

- At t = 0.567 seconds, Han shoots his second shot.

- Greedo explodes.

- Han leaves a nice tip to cover the mess he made.

Here is the real question: Did Han have enough time to react to Greedo's shot or was he going to shoot him anyway? If Han is only reacting to Greedo's failed shot, he would have made his decision and fired his blaster in just 0.32 seconds. Of course, this assumes he was already aiming at Greedo. Or maybe Han didn't react to the shot from Greedo. Maybe he was reacting to some pre-firing motion that Greedo made. But all of this has to happen, along with the firing of the blaster, in just 0.32 seconds. Let's say the movement of the firing finger and the shot take 0.2 seconds. That leaves just 0.12 seconds to react.

What is a reasonable reaction time? According to a Wikipedia article on Fast Draw[9], the best fast draw shooters can draw and shoot in 0.145 seconds. Well, Han already had his gun out and didn't need to draw it. I guess even 0.14 seconds is a possible reaction time.

Drat. I was hoping to show there was no way Han could shoot a retaliation shot at Greedo unless he was some kind of Jedi. But maybe he is. Maybe he used his hidden and unknown Jedi powers to make Greedo miss. Maybe Han doesn't even know he has these powers. Or perhaps that is just the kind of guy Han is. Perhaps this is what he was thinking:

> *"Oh, here is that Greedo guy again. I know what he is going through: pressure from Jabba the Hut to bring in someone and soon.*
> *Well, he is probably a good guy deep down. He won't shoot me, I know he won't. But you know what, I will keep my gun out just in case . . .*
>
> *What?! Did he just try to kill me? And he missed? That is it Greedo, I am going shoot you. Shooting once isn't enough for scum like you, so I am going to shoot you again. Boom. You exploded. That's odd."*

No one knows what actually happened.
Except, of course, Chewbacca.

He knows everything.

---

9   http://en.wikipedia.org/wiki/Fast_Draw

CHAPTER 5: TECHNOLOGY

CAN YOU CHARGE YOUR PHONE BY TYPING?

I like to listen to podcasts. Buzz Out Loud was one of my favorites. At one point, there was a discussion about the possibility of charging your phone simply by typing on it. The idea is that you would use the energy from your fingers hitting the screen to add more energy to the battery through some type of piezoelectric device.

Is there any way to see if this is even a realistic solution? Oh yes.

What is a piezoelectric device anyway? These things do exist. You probably come in contact with them all the time and don't even know it. Basically, a piezoelectric material produces a change in potential across it when squeezed. Why does it do that? I guess you could say that when pressure is applied to the material it becomes polarized. This polarization creates an internal electric field and, thus, a change in electric potential across the two sides. Basically, when you apply pressure, you get electricity.

Two very common uses of these piezoelectric materials are in gas grills and annoying music-playing birthday cards (which we'll get to in a minute). Have you ever wondered how the red button on your gas grill creates the spark that lights it, even though it doesn't have a little battery in there? No? Either way, when you push on that button, the pressure from your finger deforms the material inside just enough to create a spark to light the gas. You can also use piezoelectric materials as microphones, usually on acoustic guitars, as the vibration of the instrument creates electric signals.

What about those annoying birthday cards? Piezoelectric materials work the other way as well. If you apply electricity, you get pressure. If you apply a potential difference across it, you can get it to expand a little bit. If you apply an electric field, you can get the material to change its polarization, which will change its size (by just a little). I know it is much more complicated than that, but I am trying to simplify it.

For the annoying birthday cards, a very small speaker is needed. Instead of using a coil of wire and a magnet like a traditional speaker, the card has a piezoelectric speaker. A varying voltage is applied to the material causing it to expand and contract to the tune of the desired music. Maybe they aren't the best speakers, but they work. They are also thin enough that you can take them out of the card and hide them in the office next to yours. I would never do that, though.

How much energy is there in typing? Let's start by figuring out how much energy is in a typing finger. At this point I don't think it's important to find out how efficient these devices are at converting finger-pushing energy into electrical energy. I really have no idea if some materials mastermind has, or will soon, invent an awesome charging system that's nearly 100 percent efficient. For now, it's better to just assume someone could and see if that even gets in the ballpark of charging a device. Looking at a typing finger would give us an upper-limit on the amount of energy you could get from a charging device.

Let's think about how this works. Imagine a finger is in the process of pushing the screen. The screen has to compress some, even if just a little bit. If you know how much this finger moves and the force it uses, you can calculate the work done by the finger while typing.

Now, remember I am calculating the work done by the finger. What you would want is the work done on the piezoelectric device. Would these two be the same? No. The finger is going to move farther than the piezoelectric device gets compressed. The finger has to do things other than just compress the phone. But hey, let me look at the best case scenario for the charging device. What could this force and displacement be like? To experimentally estimate this I used a force sensor, which is basically a little black box with a rod coming out of it. You can use it to measure a small amount of pushing or pulling force. To simulate the finger, I put a rubber stopper on the end of the rod and began tapping the table with roughly the force you would need to type on a phone.

I found that I was pushing with a force of around 3 newtons with a displacement of around 0.0015 meters. Before moving on, let's think back to high school physics to remember how we measure force and what a newton is. One of the easiest ways to measure a force is with a spring scale. This is essentially a spring inside a tube with some markings on the side. The cool thing about springs is that the harder you pull on them, the more they will stretch. Since there is a linear relationship between force and displacement, we can get the force by looking at how far it stretches.

If you aren't familiar with the units of newtons, 1 newton is about 0.22 pounds, or about the weight of a hardback book. Why the term newton? The unit of force is named after Sir Isaac Newton, one of the people who started thinking about force and what it does to motion. There were others to work on this problem but Newton drew the long straw and had the force named after him. (I just made up that last part).

With these two values, I estimate that the work done by the finger in one tap is about 0.0045 joules. A joule is the common unit for energy used by scientists. If you want a feel for this value, take a book off the floor and put it on a table. That took about ten joules of energy. Yes, 0.0045 joules seems small, but remember that it would probably be smaller in reality. Why would it be even smaller than that? Let me point out a few things:

- I used the peak force and assumed it was over the whole distance the finger pushed. Probably not true.

- This is the work done by the finger. The work done on the device would use the distance the device is compressed. This would likely be much smaller. Think about it, the screen is less than 0.1 cm thick so it could not be compressed much.

- The device is likely to have an efficiency of much less than 100 percent.

Now we need to consider how long this tapping of a finger would take to charge a phone battery. The first question is: How much energy is stored in a phone battery? Well, one way to find out is to look online at replacement batteries for an iPhone. The first one I found had a listed capacity of 1,420 mAh with a voltage of 3.7 volts. This rating means that the battery could produce an equivalent of 1,420 milliamps at 3.7 volts for one hour.

There are two things you need to know here: First, in a circuit, the power is the product of the current and the voltage. Second, power tells you the rate of energy change.

From this, I can convert the energy rating into 1,890 joules. Since I know that each "push" is 0.0045 joules I can calculate the number of pushes to completely charge the phone. This will take 4,200,000 pushes.

So that means over four million characters need to be typed. How long would this take? How fast can you type on a phone? Just from my experience alone, it seems two characters per second would be pretty fast. This means four million characters would take two million seconds. If you typed nonstop on your phone for twelve hours a day with time left for sleeping, eating, and taking care of other bothersome tasks, it would take a month and a half to recharge your battery. That's pretty much the best case scenario. Just to be clear, though,

I could be off in many of my assumptions and the results would still be weeks and weeks of typing. Since you'd be draining your phone far faster than you'd be recharging it, this is never going to work.

Does that mean using a piezoelectric device is out? Only for charging by typing on a touch screen. There is another viable option: charging with a piezoelectric device in your shoe. Why is this different? The force exerted on your shoe is far higher than the force your finger would exert on the screen. Also, your shoe can easily compress at least 1 cm. If I start with a rough estimate of an impact force around 500 newtons, and an efficiency of 25 percent this would produce 12.5 joules per step. If you wanted to charge your phone, this would take about 150 steps, which is not too far of a distance. Of course, this assumes a rather high efficiency for these types of devices. In reality, it would likely be much lower.

What about solar charging? It seems like a great alternative for battery charging would be a solar panel. The iPhone 6 is 138.1 mm by 67 mm. What if a large portion of the back of the iPhone was a solar panel, say 80 percent? The best position for this solar panel will be if it is completely perpendicular to the sunlight.

Just knowing the size of the solar panel isn't enough. As the phone is tilted away from being completely perpendicular to the Sun, it doesn't get as much solar energy. Let me assume that the average orientation is 30 degrees off from the best angle.

In order to calculate the power from a solar panel, I need two assumptions. First, I will assume that the solar flux is about 1,000 watts per square meter. Second, I will assume the solar panel is about 25 percent efficient at converting this solar energy to electrical energy.

With all this, I get a solar power of about 1 watt if the back of the iPhone was used for charging. This seems like a much better solution than typing on the phone. If I estimate two typed characters (or any kind of push) per second, the typing would have a power of 0.009 watts. Big difference. If you could keep the phone in direct sunlight for just four hours, it would be charged. This seems plausible.

So, let's summarize. Charging your phone by typing won't work. Charging your phone by walking seems plausible, as does solar charging. Really, though, we just need better batteries in our phones.

TWEET WAVES VS. SEISMIC WAVES

If you like science and you like humor, you should check out the xkcd comics (online at http://xkcd.com). This is funny stuff, trust me. Well, xkcd is more than just funny, it's insightful as well.

One comic from some time ago (https://xkcd.com/723/) suggested that some people might read tweets about an earthquake from people closer to the epicenter before they actually felt it themselves.

People use Twitter to post things all the time. Posts about what kind of sandwich they are eating or things like "hey that was an earthquake." Of course, a tweet will travel faster than the seismic waves from the earthquake, but the earthquake gets a head start (because people don't tweet instantly and they still have to at least type "earthquake"). However, at some point, the tweet wave will pass the seismic wave.

How far away can you be from an earthquake such that someone's tweet about the earthquake gets there first? As an initial estimate, I can look at this as though it were a constant velocity problem with two things traveling at different speeds. This would be sort of like the *"one train leaves Chicago . . . "* problem, but way cooler.

To do this problem I need some starting estimations. Here are the values suggested by the xkcd comic (is it even considered a comic?):

- The speed of the seismic wave is about 3-5 km/s (I will call this v_s).

- There is a time delay between the earthquake and the first tweet response. Let me call this t_t and estimate it at around 20–30 seconds.

- How fast does the tweet move? This will be v_t and I will start with a value of 200,000 km/s.

Now for a little bit of math. If you want the position of the leading edge of the seismic wave (the "tweet wave") I can say that the position of the seismic wave is the velocity of the wave multiplied by the time (you know, your basic distance equals rate times time). The tweet wave position function will look similar except that it has a different velocity and starts at a later time.

By solving for the time the two waves are at the same position, you can find out how far away you need to be to get a tweet about an earthquake before feeling it.

One way to solve this problem is to plot both of these functions and see where they cross. This is what you get:

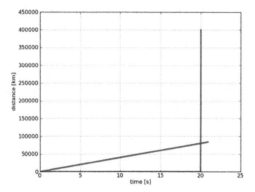

Here you can see the problem. The tweet-speed is so much faster than the seismic wave speed that it only matters how far the seismic wave has traveled in time delay. Anyway, with this model the seismic wave travels 80 km during that delay. The recent earthquake centered in Mineral, VA would look like this:

Anyone outside that circle would have a chance of getting a tweet warning of an oncoming earthquake.

Back to the seismic wave. Is there any way to estimate the speed of a seismic wave without just plain guessing it (or looking it up)? It turns out there might be a way. Along with this 2011 Virginia earthquake, there was a video showing the different detectors across the United States. (In case you want to see, here it is: http://youtu.be/lKE7MLNdtcg)

With this video, I can use the basic tools in video analysis to get the position of the leading edge of the disturbance as a function of time with a wave speed of about 7 km/s.

Now, what about the speed of a tweet wave? Let me estimate this speed with a simple experiment. Suppose I post a note on twitter and see how fast the responses to this post are. Here is my tweet:

By looking at the difference in the time between the original tweet and when the person received the tweet I could get the speed (since I know their location). It seems like a simple experiment, right? Well, there was one small problem. I sent the tweet at 1:48 p.m. Central Time. Most of the kind people who responded reported receiving my message at 1:48 p.m. Some even reported a time of 1:47 p.m. So, it seems our clocks are not synchronized. On top of that, the wave's travel time was smaller than one minute.

However, some of the responders were quite far away. I had one in Germany (about 8,200 km away) and one in South Africa (13,000 km). Using this, I would have a lower limit for the tweet-speed at 217 km/s.

This is a speed much smaller than my first starting value of 200,000 km/s. Perhaps I need another experiment. With the help of my brother, I measured the time between my post and his viewing. For him, I found a tweet speed of 35 km/s. So maybe a speed of 100 km/s seems okay.

Now with this new data on both the tweet and seismic wave speeds, I can make a new plot. Let me assume a response time of one minute. This gives the following plot:

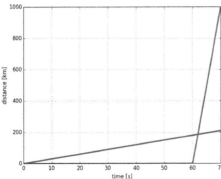

Here the steeper line is the leading edge of the tweet wave.

This would make a new circle around the earthquake with a radius of around 180 km. People outside this circle are far enough away from the center of the earthquake that they didn't even need a warning. So, it seems Twitter won't give you a good warning for an approaching seismic wave.

HOW REALISTIC IS *ANGRY BIRDS*™ PHYSICS?

What does a force do to an object? What is a force, for that matter? A force is an interaction between two objects. At the most fundamental level, there are only four types of forces. There is the gravitational force, which is an interaction between objects that have mass. Next is the electromagnetic force, which is an interaction between objects with electric charge. The last two are the strong and weak nuclear forces. The strong force is a short-range interaction between particles like the proton and neutron. The weak nuclear force is another force between sub-atomic particles. The last two don't come into play in *Angry Birds*™, so let me just leave it at that.

What about forces like the force of a ball pushing on my head after someone threw it at me? What kind of force is that? Well, technically, that is the electromagnetic force. The atoms in the ball and the atoms in my head both have electrons and protons. When they get close, there is an electromagnetic interaction between these charged particles. It feels like I was hit by the ball. But was I really? It is hard to define "touch" at the atomic level. Let's just say there was an interaction between the atoms in the ball and the atoms in my head.

Okay, I'm getting off track. What about the gravitational force? Any objects that have mass will interact with any other objects with mass. However, the interaction is very weak. Typically, we don't even notice these gravitational interactions unless at least one of the objects has a very large mass.

Let me give you an example. Suppose I hold a pencil and let go. The pencil has mass and so does the Earth. I do, too. The pencil's mass and my mass aren't huge. This means that there is a gravitational interaction between the pencil and me, but it is extremely small. It isn't large enough to measure or to notice any motion due to this interaction. But the mass of the Earth is huge. This means the gravitational force from the Earth is large enough to affect the motion of the pencil. The pencil then falls.

Fine, but what do forces do to objects? In short, forces change the motion of an object. Change is important here in case you can't tell. Don't fall into the trap that most people fall into. The common idea is to say that forces cause motion. In a sense, they could be correct, but for the most part they are mistaken. Here is a great question to test your family and friends: what would happen if you have one single constant force on an object? I bet that 96.7 percent of the responders would say, "Constant force makes an object move at a constant speed."

Sadly, the idea of constant force resulting in constant motion is totally wrong. I mostly blame Aristotle. The reason everyone believes this is because it mostly works. The problem is that it is very difficult to just get one force acting on an object in real life. If you did manage to do this, you would find that a constant force constantly changes the motion of the object. If this force was in just one direction on an object that was initially at rest, the object would just keep increasing in speed.

How much does this motion change? Let me go ahead and call the "change in motion" the acceleration. The acceleration of an object depends on two things. It depends on the strength of the force and the mass of the object. The greater the force, the larger the acceleration. The greater the mass of the object, the smaller the acceleration. You probably know of this as Newton's second law which says that the Net Force = Mass x Acceleration.

One more thing about forces: remember the falling pencil from before? If the pencil has mass and the Earth has mass, wouldn't the pencil also exert a force on the Earth? Yes, it would also exert a force on the Earth. This force on the Earth would be the same magnitude as the force the Earth exerts on the pencil. But there is one big difference: the mass. Like I said before, the mass of the Earth is huge. This means that in order to have a significant acceleration, the force would also need to be huge. So, although the forces on the pencil and the Earth have the same magnitude, one has a significant effect and the other doesn't.

Now, that's how physics works in our world. Let's look at Angry Birds™ to see if it works like our world. Wait, you still don't know anything about the game Angry Birds™? Please, tell me this isn't true. Alas, let me give a brief description of this very popular game.

Essentially, the goal of the game is to use a slingshot to launch birds at structures housing pigs. The structures protect the pigs and can be made of glass, wood, or stone. Why do you shoot at pigs? Why are the birds angry? Why can't they just fly? Sadly, I don't know the answers to these questions. All I know is that the game is strangely addicting. You can play it on almost all smartphones as well as on PC (or even in the Google Chrome web browser).

I guess I should say a couple more things. In *Angry Birds™* there are different kinds of birds that you can launch. You don't get to pick which birds to use, that is decided for you. However, different birds can do different things.

If you start *Angry Birds™*, the first level gives you the red bird (I will call him Red). If you shoot Red, what happens? Red moves through the air with a para-bolic path. It turns out this is exactly what should happen for a physics motion that we call projectile motion. Why does it move like this?

There is only one meaningful force acting on Red after he leaves the sling-shot. Trust me on this one. There is not a "force from the slingshot moving Red forward." I know you want to add that other force, but don't. That would be Aristotle talking to you. Don't listen to him. Why is there only one force? Think of it this way: you can break all the forces into two categories. There are forces that are the result of things "touching" the object and forces from objects that don't have to touch—long range forces. What is touching Red after he leaves the slingshot? I guess you could say "the air" as a valid answer, but let's just pretend that the forces from the air are small enough to be neglected in this case. What else? Nothing else is touching Red. Okay then, what about long range forces? There is just the gravitational force. Red flying through the air is what happens after the slingshot's application of force.

With just the gravitational force acting on the bird, I should say something about gravity. On the surface of the Earth, we can model the gravitational force as having a constant magnitude and directed straight down. The magnitude of this force would be the product of the mass of the object and the gravitational field, which has a value of 9.8 newtons per kilogram.

Because this force is only in the vertical direction, it only changes the motion of the bird in the vertical direction. What does that say about the horizontal direction? Well, if it doesn't change its horizontal motion it must be moving with a constant speed in the horizontal direction. This is true for physics-book pro-jectile motion, but what about *Angry Birds™*?

How could you even measure the horizontal motion of Red? Well, there are lots of ways. You could just make a video of the bird in motion, either by using screen capture or pointing a video camera at your game. Then you could see how far horizontally it moves in each frame. There is another option as well. You could use video analysis software. Just as I stated in the previous chap-ter, the basic idea is that you can mark the location of an object in each frame of the video and get x, y, and time data from its motion. There are several software packages that do this. Many classes use the Logger Pro software from Vernier. Personally, I prefer the free (and more powerful) Tracker Video from Doug Brown[10].

A plot of the horizontal position for Red after he is launched from the sling-shot says that during each time step (frame), he moved horizontally the same amount. Or perhaps you could say the horizontal velocity is constant.

In this case, it is 2.46 distance units per second. But what are the distance units? They aren't really in units of meters. In this case, they are in *Angry Birds*™ units (I will call them AB instead of m). When setting up a video analysis, I have to tell the program how big things are. That is, each pixel on the screen corresponds to real distance. Unfortunately, I don't know what that real distance is. So, I defined the AB unit to be the length of the slingshot.

Can there be a way to relate this *Angry Birds*™ distance to a real life distance? Maybe so. Let me do this by looking at the vertical motion of the bird. Gravity is the only force on the bird and it is constant. This means the vertical acceleration of the bird will also be constant. A constant vertical acceleration would mean the vertical velocity changes at a constant rate. What does a graph look like with a constantly changing slope? A parabola. Here is the plot of vertical position versus time for Red:

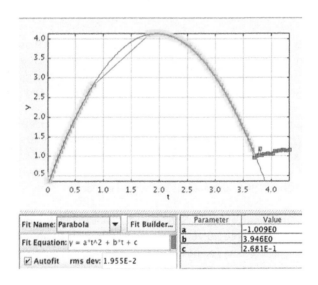

I won't go into all the details of where this comes from, but if you have an object with a constant acceleration then its position should have the following function:

$$y = y_0 + v_{y0}t + \tfrac{1}{2}at^2$$

Since Tracker Video was kind enough to fit a parabolic function into the data, I can see that the term in front of the t^2 is -1.0 AB/s² (remember AB is the unit of distance in *Angry Birds*™). This would give a vertical acceleration of -2 AB/s².

Now here is the cool trick. What if I assume that *Angry Birds*™ takes place on Earth? In that case, the vertical acceleration should be -9.8 m/s² for an object that is only acted on by the gravitational force. If I set these two values to be equivalent, I can get the relationship between the units of AB and meters. Doing this simple conversion, I find that one AB equals 4.9 meters. The sling-shot is almost 5 meters tall. What about Red? Using the same scale, Red is almost 70 cm tall. That is one big bird. A big, angry bird.

Let's take a look at the blue birds. Of course, the blue birds need a name. I will call them the Blues. If you aren't familiar with the Blues, they have a special ability. When in the air, the Blues can split into three other blue birds. Fairly useful, right? Also, the blue birds seem to be adept at breaking glass structures more so than any other bird.

So, here is the question: What happens to the mass of the Blues during its transformation into three birds? Is the mass of each of the three new birds one-third of the original mass? Perhaps each blue bird has the same mass as the original (which would mean that mass is not conserved in *Angry Birds*™).

How will I measure mass in *Angry Birds*™? Is there a special scale? No, but that is okay. This is one of the cool things about games such as *Angry Birds*™. You want to try and answer a question but you don't have everything you need. That means you have to come up with some alternative experimental evidence. This is just how science works in the real world.

Let me give a real world example. Suppose you want to find the mass of something like the electron. Could you just use a balance or scale and move on? No, of course not. Instead you have to find some other way to determine the mass. In fact, for this case you can first determine the mass-to-charge ratio for an electron and couple that with an experiment to determine the charge. Complicated, I know, but that is the way science happens and what makes it so much fun. Would Mount Everest be fun to climb if it was easy? No, it wouldn't be the same if it were easy.

Back to *Angry Birds*™. How can we look at mass? What actions should depend on the mass? What about collisions? Okay, first a crash course in collisions. Suppose I have two objects interacting (object A and object B). During the interaction, object A pushes on object B, and B pushes back on A. These two forces are the same magnitude. It doesn't matter what their masses are.

So, the forces are the same (well, the same magnitude). But what does a force do? Forces change momentum. There are two important points about momentum to keep in mind (although I could spend an entire semester on momentum): momentum is the product of mass and velocity for that object. Momentum is also a vector (meaning the direction matters).

During this collision, the two masses have the same forces for the same time. This means they have the same (but opposite direction) change in momentum. Don't forget the "change in" part, it's important. The point is that if I can look at the change in velocity for the objects, I can infer ideas about their masses, assuming momentum is conserved.

Now for an experiment in *Angry Birds*™. All I need to do is find a level that has a suitable target for me to shoot the Blues at. If you look carefully, you can find a level with a rock sitting on a short shelf. For example, take a look at level 3-3 in the original *Angry Birds*™. This rock is sort of like a baseball on a stand in tee-ball. I can shoot the Blues at this rock and use video analysis to measure the post-collision speed of both the bird and the rock.

By measuring the speeds of the objects after the collision, I can get a ratio of the mass of the objects (assuming momentum is conserved). Let me start with Red, just for comparison. Let me show you one example of how I would get these velocities both before and after the collision.

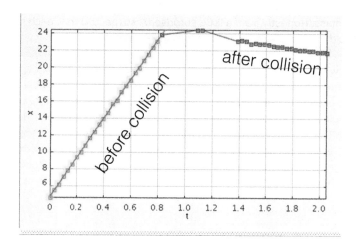

This is just the velocity in the horizontal direction. The collision taking place (at least the interaction forces) are in the x-direction, so I don't really need to worry about the vertical (y-direction) motion. This is a good thing since momentum isn't conserved in the vertical direction because of the gravitational interaction.

After finding the velocity of Red before and after the collision, I can find the velocity of the rock after the collision using the same method. If momentum is conserved then the mass of Red is 0.31 times the mass of the rock. Yes, I know I left out a lot of the algebraic details. I assume you will forgive me for this. Maybe I should make Red the standard unit of mass in *Angry Birds*™. In that case, the mass of Red would be 1 mr (where mr is the Angry Bird mass unit, standing for mass of Red) and the rock would have a mass of 3.1 mr. It's the best I can do. There is nothing in the game to connect the mass of the bird to masses in real life.

What about the Blues? Here, I need to do two things. First, I will shoot the Blues at the same rock without expanding it into three birds. After that, I can expand the Blues so that one of the three new blue birds hits the rock.

After shooting the single (unexpanded) Blues at the rock, I find that it has a mass of just 0.019 times the mass of Red (0.019 mr). What a tiny baby bird. Red has a mass sixty times larger.

Now, if I split the Blues into three birds and collide with the rock I can also get its mass. In doing this, I find it has a mass of 0.29 mr.

Is this odd? I guess it's a little bit odd. This newly created blue bird has a mass about fifteen times larger than the original bird. If you take all three newly created blue birds, the total mass after the "expansion" is forty-five times the original mass. Honestly, I am a little surprised. I suspected that each new blue bird would have the same mass as the original Blues. Weird.

Okay, how can I explain this lack of conservation of mass? Yes, I know that it is a video game and not real life, but I am going to try and make it work anyway. The first explanation is that momentum is not conserved in *Angry Birds*™ and why should it be? It doesn't have to follow the same rules as the world we live in, right? The second explanation is that when you tap the screen, you open a portal to another dimension. The original Blues then goes into another world exchanging places with these three new birds. These new birds look like the one but they aren't the same. They are made of titanium or something.

What does it all mean? It means you can gain a big advantage by using "tapped" blue birds. You would be a fool to leave the Blues as they were. Tap away!

Before looking at the physics of the yellow bird,
maybe we should pause for a haiku:

> *The Sun and sky, still.*
> *Green pigs cackle with delight.*
> *Smash wood from above.*

Oh sure, there are debates about the proper form for a haiku. But there is no debate about what the yellow bird does. First, it can smash through the wood blocks better than most birds. Second, when you tap the screen while the yellow bird is in motion, something happens, but what exactly happens?

When I first started playing *Angry Birds*™ oh so many years ago, I had this feeling that after tapping the screen the yellow bird went at a constant velocity. Then one day, I accidentally shot it super-high. The bird didn't just keep going up, it looked like parabolic motion. I guess I was wrong. What next? Collect some data.

Right before you tap the yellow bird, it has some velocity. Clearly, it has a different velocity after the tap. Maybe there is some constant tap-acceleration in the direction the bird was moving.

After doing a video analysis on the yellow bird, you can get more insight into what actually happens during (and after) the "tap." A plot of both the horizontal and vertical velocities shows that the speed of the bird increases after the tap. Also, there is no horizontal acceleration before or after the tap (only during the short tap time). For the vertical motion, the acceleration during the non-tap time seems to be constant just like it was plain projectile motion.

It still is not clear what happens during the "tap." If you look at several different yellow bird shots, one thing does seem clear: The tap has some type of effect that lasts 0.067 seconds. During this short time, the total acceleration will be due to the changes in both the x- and y-velocities.

After quite a few shots, it seems that the accelerations for all of these shots are not the same. Instead, it almost appears as if the acceleration during the tap is whatever it needs to be to get the yellow bird's speed up to some post-tap value of about 30 m/s.

Notice that I said almost. It doesn't always work this way. Strangely, in some cases, the yellow bird behaves quite differently after the tap. If you tap the yellow bird after it is already moving down, it doesn't have the same vertical acceleration as the other yellow bird shots. Here is a plot of the post-tap vertical acceleration for birds versus the angle the bird is moving at the time of the tap:

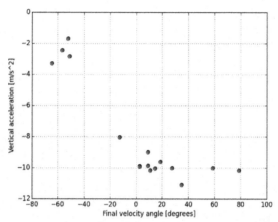

Notice that most of these birds have an acceleration value around -9.8 m/s² as you would expect. However, if the bird is moving at a downward angle around 50 to 60 degrees below the horizontal, the vertical acceleration is only around -3 m/s². Weird. I guess the game developers felt the yellow bird would end up going too fast if the angle was too steep. It is their game and thus their rules.

This example shows how you would figure out the rules in the *Angry Birds*™ game. Sometimes the answer is clear, but sometimes you just get more questions.

IPADS ON A PLANE

There was a recent news story about an airline using iPads in its plane. The short version is that American Airlines obtained FAA approval to replace its huge flight manuals with iPads. As far as I understand, the flight manual contains everything you might possibly want to know about the aircraft and weighs 35 pounds. American Airlines claims that by switching from paper manuals to iPads, it will save 1.2 million dollars in fuel over the course of a year.

First, how do airplanes fly? Just a quick note: this is a much-debated topic. So, that is your warning. Let me give an extremely simple explanation of flying. This will be so simple it will probably make you angry.

Imagine an airplane wing moving through the air. Don't imagine air as some continuous material. Instead, think of it as several tiny balls. What about the wing? Let's say the wing is just a tilted board. Yes, the wing has to be angled up, even if just a tiny bit. This slanted board is moving through the tiny balls of air and colliding with them.

When one of these balls collides with the wing, it changes its momentum. With a change in momentum, there must be a net force. The net force for this air-ball's change in momentum comes from the wing. Because forces are an inter-action between objects, the air must also exert a force on the wing with the same magnitude. When a number of these air-balls collide with the wing, the net result is a "lift" as well as a "drag."

Of course, there has to be another force. If you want your plane to go at a con-stant speed you would need a thrust force to balance the drag force. But what about the weight of the plane? If you increase the weight, you would also have to increase the lift. How can you do this? There are two ways. First you could have the plane go faster. This means you get more collisions with air-balls as well as a greater change in momentum for the air-balls. Both of these would lead to greater lift.

The other option would be to change the angle of the wing. If the wing is tilted more towards a vertical orientation, air-balls will be deflected down more and result in a greater upward force. Of course, this would also produce more drag. So either way, more lift means more drag. There is no escaping this effect.

"But what about Bernoulli's principle?!" Yes, you are correct. I didn't say any-thing about pressure and lift or anything like that. Why not? Well, if you treat the air as particles, you don't need to deal with pressure and Bernoulli's prin-ciple. You would use that if you wanted to treat the air as a continuous fluid. Even so, the air-ball collision model isn't perfect. However, it is useful enough to show how increased mass in a plane would lead to more drag and thus require more fuel.

Before looking at the fuel, let me first look at the money. The claim by American Airlines is that replacing a 35-pound flight manual with an iPad will save $1.2 million per year. Now, I am not sure how many flight manuals this would replace, but I suspect it would have to be replaced by more than one iPad for redundancy. If one iPad weighs about 1.5 pounds, then the net weight savings would be 32 pounds (14.5 kg). From this, I get a yearly cost savings of $82,000 per kilogram of reduced mass.

I can use this yearly savings per kilogram to look at other decreases in mass. This isn't a bad assumption because if the change in mass is tiny compared to the total mass of the plane, then it will at least be approximately true.

What about one bag of peanuts? Suppose that each American Airlines flight removed just one bag of peanuts (that they no longer give you) from the flight. How much in yearly fuel savings would this be? Well, first I need the mass of a bag of peanuts. Instead of searching for the exact mass of a bag of peanuts, let me just get a ballpark mass of 25 grams. That seems close. So if each American Airlines flight reduced the payload mass by 25 grams, they would save $2,069 per year.

That is $2,000 a year for one bag of peanuts per flight. What if they took away all the peanuts? Here I will need to make a quick guess. How many bags of peanuts would be brought on board each flight? This is tough because I want an average. How about an average of three-hundred seats per flight with an average four-hundred bags of peanuts? Sounds good enough for me. This would lead to a yearly cost savings of right around $800,000! That is quite close to the savings from the paper-based flight manual. As a bonus, the airline would also save money by not buying peanuts.

What about luggage? Like most airlines, American Airlines charges a fee for you to check a bag on the plane. According to American Airlines[11], the bag can have a mass up to 23 kg without another extra fee. So, how does the fee they charge the passenger compare to the price of fuel needed to accommodate this extra mass? Ok, this will be a little more difficult. Of course, that won't stop me.

Here are the fees according to American Airlines:

- Flight within the US: $25 for the first bag.

- Flight in the US: $35 for the second bag.

- Flight in the US: $150 for more bags. Wow, that seems harsh.

- Flying from the US and going through Europe makes the second bag $60. What if you go around Europe instead?

---

11   https://www.aa.com/i18n/travelInformation/baggage/baggageAllowance.jsp

I am going to stop there. It seems like $25 for the first bag is the price for all "first" checked bags. Now I need to make some estimates. What is the average number of checked bags per flight? What is the average checked bag mass? Let me give a range of values.

How about an average checked bag mass of around 9 to 14 kilograms? Remember, I am just guessing here. What about the average number of checked bags per flight? If I stick with three-hundred seats per flight then maybe a little less than half of these passengers check a bag. I will estimate one hundred twenty to one hundred eighty checked bags. Oh, and these are "first checked" bags.

Now I can estimate the mass of these "first checked" bags at somewhere between 1,080 to 2,520 kilograms per flight. With this I can estimate the fuel costs. Based on the estimate from above, it would cost the airlines about $89–$208 million a year.

So is the $25 fee excessive? This is a bit tougher to answer. The above calculation just looks at the yearly averages. It doesn't look at the cost per bag. I would need to know how many people are paying for checked bags a year. According to an American Airlines information page[12] an average day has 275,000 passengers with an average 3,400 flights and more than 300,000 pieces of luggage.

With that estimate, I get an average of just eighty passengers per flight. I suppose my first guess wasn't that great. But that number also says there would be 1.09 bags per passenger. If I go with eighty passengers per flight and half a bag per passenger, this would change the fuel costs due to luggage to $30–$46 million per year.

How much money does the airline collect in terms of baggage fees? Suppose that only half of these passengers pay the $25 baggage fee. In that case, American Airlines would collect $1.25 billion in fees. That is quite a bit more than my estimated fuel costs.

Maybe the fee is there to cover all the bags, including the carry-on bags. Let's say these 100 million passengers per year bring an average total luggage mass of 30 kilograms. The yearly fuel cost for these bags would be $198 million. This is still much less than the $25 baggage fee.

What would a fair baggage fee be? Well, first you have to choose what this fee

---

12 http://www.aa.com/i18n/amrcorp/corporateInformation/facts/amr.jsp

pays for. The airline has lots of expenses related to luggage. There are the people who handle them, the space on the plane, and other considerations. Let's go back to the first estimate for the yearly cost of checked bags at $46 million per year. If I stick with the 50 percent of passengers checking a bag on a flight, this would be 50 million passengers. This makes the calculation pretty easy. One dollar per checked bag. Okay, change it to two dollars per checked bag. I wouldn't mind paying two dollars per bag.

One important point: The baggage fee calculation might not be valid. It is a slight stretch to assume that change in fuel costs for a mass difference of 14.5 kg is the same linear function for a mass change of 2,000 kg. Even though the fuel-mass function isn't likely to be linear the whole way, it probably still gives a fair estimate. Remember, this is just an estimate but I still think two dollars a bag would be reasonable.

Okay, but back to the fuel savings. If American Airlines can save $1.2 million per year with the lower mass of the iPads for flight manuals, are there any other ways to shave off some weight during the flight? Yes, and I have an idea how. What if all passengers had to relieve themselves of excess bodily fluids before the flight? Just to be clear, I am talking about urination.

Let's go back to the eighty passengers per flight. Suppose each one stops by the restroom before getting on the plane. How much mass would this account for? I will guess that on average, a person could produce 300 ml of fluid. Oh sure, some people could fill a whole liter, but there could be some passengers that have stage fright and can't pee during a flight. I think 300 ml is a fair, if low-end, estimate. It seems plausible that urine would have a density similar to that of water (although I have not experimentally verified this).

Using a urine density of 1,000 kg/m^3, I get an average urine mass of 0.3 kg. With a flight of eighty passengers, this is total mass savings of 24 kilograms. Using the same model for the savings from the iPad, this would have a yearly fuel savings of $1.98 million. Getting passengers to visit the toilet before the flight would have the added benefit of not having to get up in the middle of the flight to visit the restroom. It makes sense and saves money. This should be a law.

CHAPTER 6: SPORTS PHYSICS

DIVING AND THE MOMENT OF INERTIA

Let's look at the 10-meter diving event in which divers leap from a platform 10 meters above the water. Scoring is based on several factors, including the height and difficulty of the dive, but I will focus only on rotations. Let's look at how a diver rotates and what matters in a rotation.

First, how long does a 10-meter platform dive last? This isn't too difficult of a question. If we assume the vertical acceleration of the diver is constant, then we can use the typical kinematic equation that relates position and time for an accelerating object. Using a vertical acceleration of 9.8 meters per second squared gives us a fall time of 1.42 seconds. A 10-meter dive happens fairly quickly.

What about angular momentum? The thing most people don't realize is that once the diver starts to fall, the angular momentum essentially stays constant. What is angular momentum? Maybe we should look at linear momentum first (usually just called "momentum").

The magnitude of momentum is the product of an object's mass and velocity. I say "magnitude" because momentum is a vector such that direction matters. To make things simpler, I will assume we are only dealing with a change in magnitude. So, how do you change the magnitude of the momentum of an object? In short, a change in momentum is due to a net force on an object. If the net force on an object is zero newtons, then the momentum doesn't change. Again, change is the key word here. If we use this principle for a falling diver then the vertical force does change the momentum. As the diver falls, the momentum of the diver increases.

Now, what about angular momentum? In one sense, angular momentum is just like linear momentum except that it deals with rotational motion. Perhaps it would be better to call this the "rotational momentum." Angular momentum (I will refer to its traditional name) also depends on two things: the angular velocity and the moment of inertia.

The angular velocity is fairly easy to understand since it is just a measure of how fast the object is spinning. But what about the moment of inertia? It might make more sense to call this the "rotational mass." It is the property of the object that makes it more difficult to change the object's angular velocity. How do you change the angular momentum? Instead of a net force, you need a net torque.

Torque is different than force. I don't want to talk about torque too much except to say there is no torque on the diver after the diver leaves the platform. Although there is a gravitational force on the diver, it doesn't cause a rotation.

Let me explain my favorite moment of an inertia demo—a demo you can do on your own. For this example, I have two sticks of some type with some masses attached to them. I used juice boxes taped to PVC pipes.

On one pipe, the two juice boxes are near the center of the pipe while the other places them near the ends. Both objects have almost the exact same mass. However, if you hold the center of the pipes and try to change the rotational motion (twist them back and forth), you will find the pipe with the juice boxes on the ends is much harder to rotate back and forth.

So, the moment of inertia not only depends on the mass but also the location of this mass relative to the point of rotation. The further away the mass is from the rotation point, the greater the moment of inertia.

What does this have to do with a diver? During the jump, the diver has to push off the platform in such a way that there is a torque to change the angular momentum from zero to something. This also gives the diver some rotational motion. Now, suppose the diver wants to do a triple tuck? How can the diver do this in less than two seconds? You can't change the angular momentum, but you can change the moment of inertia.

By pulling the legs and arms closer to the point of rotation, the moment of inertia decreases and the angular velocity increases. A tighter tuck means a faster rotation. But how do you stop rotating to enter the water? You don't stop rotating because you can't. The best you can do is to straighten your body again and increase the moment of inertia again to decrease the angular speed. Yes, it's a tough move, but that's why they are Olympic divers.

CAN A HUMAN OUT-PULL A TRUCK?

There are many shows on TV. Some of them claim to be about science, but perhaps they really aren't. One such show goes by the name *Sport Science*. Obviously, I am not a huge fan. It has great graphics but is lacking in the "science" part of the sport.

In one episode of *Sport Science*, the power of NFL running back Marshawn Lynch is compared to the power of a truck. You can't watch a video in a book, so let me briefly describe the series of events in this particular show.

- Marshawn Lynch is cool. (Likely true.)

- Marshawn is equipped with wireless motion sensors to measure his every move and create a real-time animated skeleton of his body. Other than this animation, the sensors have no purpose.

- Marshawn pulls a 585-pound sled on some AstroTurf over a distance of five yards in about eleven seconds. They use these values to calculate Marshawn's power.

- Then they get a 6,700-pound diesel truck with a 325 horsepower engine to pull 17,000 pounds of concrete barriers on asphalt. Just to be clear, the ratio of the weight of Marshawn to the weight of sled is the same ratio for the truck and the blocks.

- Perhaps not surprisingly, the truck doesn't pull the blocks. Instead, just one wheel spins (they didn't even use a four-wheel-drive truck).

Ok, that is the episode.

There are two things that are not quite right with this episode. First, the power thing. How do you calculate power? Power is the rate at which work is done. So, just take the work the person does and divide by the time it takes to do this work. If a person is pulling something, the work done will be the force required to pull the thing multiplied by the distance the thing moves (assuming the force is in the direction of motion).

Sport Science claims Marshawn produces 573 watts per kilogram. How do they get this number? At first, I thought they took the weight of the sled and tires (2,600 newtons) times the distance (4.6 meters) and divided by the time of eleven seconds. This gives a power of about 1,000 watts or, since Marshawn has a mass of around 100 kg, about 10 watts per kilogram.

Odd. This doesn't even come close to their value of 573 watts per kilogram. The total power (according to *Sport Science*) would be 57,000 watts. That is crazy huge. He'd have to pull the weight across two football fields in eleven seconds to get near that.

Even if Marshawn pulls the sled 4.6 meters in five seconds (I am trying to help with a smaller time), he would have to pull with a force of 62,000 newtons (14,000 pounds). I am sure Marshawn is a strong dude, but not that strong. I have no idea where they got this value for power.

While I am all worked up, let me point out another error. The work depends on the force Marshawn exerts on the sled. It doesn't depend on the weight of the sled. Anyone can move 585 pounds. In fact, I had my six-year-old daughter pull the family car. She's strong for her age, but it isn't that difficult if the car is on level ground. Just make sure an adult is in the car to apply the brake to keep the car from rolling away.

If there isn't much friction, even a small force can get the object to change its motion. It just takes a little bit longer to speed up. You don't use the weight of the car to calculate the work (and thus the power) unless you are lifting it straight up.

Now, what about friction? Let's go back to Marshawn pulling a sled with tires. Suppose for an instant that he pulls at constant speed (which is not such a bad assumption). In this case, the sum of the forces on the sled must be zero. If they weren't, the sled would accelerate. So what forces are acting on the sled? Well, there is the gravitational force pulling down on the sled and the force of the ground pushing up. If there were no other forces, these two would be equal in magnitude. Finally, there is the frictional force in the opposite direction that Marshawn pulls.

If Marshawn pulls at an angle, things are a little more complicated. Let me just assume that he pulls horizontal to the floor. In this case, he pulls with a force equal in magnitude to the frictional force on the sled. The frictional force on the sled depends on both materials rubbing together (metal sled and AstroTurf) as well as the force of the floor pushing up on the sled.

What about the forces on Marshawn? Whatever force he pulls on the sled will also pull back on him. Since the net force must also be zero, the frictional force on him must be the same amount he pulls on the sled. Remember, this is for the case of a horizontal rope. If he pulls upwards a little bit on the rope, it will increase the force the floor pushes up on him which will in turn increase the frictional force. At the same time, this will decrease the force the floor pushes up on the sled and decrease its frictional force.

You also have to consider the different types of surfaces between the floor and the sled as well as the shoes and the floor. If Marshawn was wearing dress shoes with leather soles, he wouldn't be able to pull the sled no matter how strong he is. Instead, he would just slip. If the sled bottom was covered in the same cleat material as real shoes, he would slip as well. This is because the frictional force on the sled would be greater than he could ever pull on the rope since the sled has a greater mass.

On *Sport Science*, they were trying to compare the power that Marshawn had to a truck. To make things fair, they wanted something similar to a person pulling tires—this was 2.6 times Marshawn's body weight. They used a 6,700-pound truck. 2.6 times its weight is around 17,000 pounds. That's fair, right? Ah ha! It is not fair. First, the truck was on asphalt and it was pulling some concrete barriers on asphalt. Second, the tires were spinning. How can that be fair?

You don't even need to pull anything to get the power per kilogram for the truck. They say in the video that the truck has a 325 horsepower engine. With the mass of the truck, this is equivalent to 80 watts per kilogram.

In the end, the episode tried to show that Marshawn was more powerful (per kilogram) than a truck. This likely could be true in short distances. Since he has a mass of 100 kg, he would need to produce more than 800 watts to win. This would be difficult, but maybe not impossible. If he pulled a sled 4.6 meters in ten seconds, he would have to pull with a force of 1,739 newtons (390 pounds). That's tough, but possible.

You know what? You could actually measure the force that he pulls on the sled. All you would need to do is put a spring scale in between him and the sled. That, along with a stopwatch, would give you all that you needed. Of course, you wouldn't get the impressive skeleton-animated graphics, but you could do some cool science.

THE LONG JUMP: GRAVITY AND AIR

Even now, there are those who claim the long jump record of 8.9 meters that Bob Beamon set in 1968 was only so awesome because he accomplished it in Mexico City, which is almost 8,000 feet above sea level. The argument is that the air is thinner, meaning there is less air resistance. Also, Mexico City is further from the center of the Earth, therefore the gravitational forces are smaller. Does any of this have any impact? If so, does it matter?

First, let's look at gravity. On the surface of the Earth, the usual model for gravitational force is the object's mass times the gravitational field (g) where g is about 9.8 newtons per kilogram. So, a 1 kg object would have a gravitational force of 9.8 newtons (directed down).

However, this model doesn't work if you get too far from the surface. The gravitational force is an interaction between two objects with mass, and the magnitude of this force decreases as the two objects get farther away. This is often referred to as the Universal Law of Gravity since it should work everywhere. Everywhere in the universe, that is. In it, the gravitational force is proportional to the product of the two masses interacting and inversely proportional to the square of the distance between the objects.

If you use the universal gravitational force on the surface of the Earth, you would put the mass of the Earth as one of the masses and the radius of the Earth as the distance between these two things. It shouldn't be surprising that you would mathematically get a value of 9.8 newtons per kilogram. This is the same value as the other method for calculating the gravitational force. Wouldn't it be weird if these two models gave different results?

What if you aren't near the surface of the Earth? What if you were in Mexico City with an elevation of 2,240 meters above sea level? With that value for altitude, an object would have weight that is 99.93 percent the weight of the object at sea level. Not a big difference, no. But is it a big enough difference to mean a new world record long jump?

The above comparison of weights at sea level and at elevation would be valid if gravity was all that mattered. In terms of the apparent gravitational force, there are two other issues. First, the Earth is not a uniform sphere with uniform density. If you are near a mountain, the mass of that mountain can affect the gravitational field in the area, even if you are at sea level.

The second consideration is the rotation of the Earth. The closer a location is to the equator, the faster that location has to move in a circle as the Earth rotates each day. Mexico City is about 19.5 degrees above the equator, so it must move fairly fast. Of course, if you move in a circle you aren't in a non-accelerating reference frame. In order to treat it like a stationary frame (which is what it seems like to us on the surface), you would have to add a fake force called a centrifugal force pointing away from the axis of rotation. The combination of centrifugal force and the actual gravitational force would be the apparent weight.

If Mexico City were at sea level, this rotational motion would cause the apparent weight to be 99.69 percent the value of the Earth without the rotational effect. Putting both the gravitational and rotational effects together,

the apparent weight at the elevation of Mexico City would be 99.62 percent the value at sea level without rotation. So that's not much difference.

Okay, so the gravitational differences and rotational effects don't seem to be very large. What about other factors? What about the density of air? As you move farther away from the surface of the Earth, the density of the atmosphere decreases. With a lower density of air, there would be less air resistance on the jumper during the motion.

The commonly used model for the force of air on a moving object has it as proportional to the square of the object's speed and proportional to the density of the air. If you decrease the density by a factor of two for the same speed of a moving object, you would have half the air resistance force.

It turns out that the density of air isn't the most straight-forward thing to model. This would seem to fall under the same category as "weather" when it comes to tough stuff to predict. Needless to say, there is a simple model to use[13].

Using the referenced model, I find that at sea level the density of air is about 1.22 kilograms per cubic meter compared to 0.98 kilograms per cubic meter at an elevation of 2,240 meters. Would this decrease in density have as much of an impact on the jump as the decrease in the gravitational force?

---

13   You can find more details about modeling the density of air on this Wikipedia page. http://en.wikipedia.org/wiki/Density_of_air#Altitude.

The motion of an object moving through the air with air resistance isn't a simple problem. What makes it so complicated? Without the air resistance, the acceleration of the object would be constant. If you think about plain projectile motion, an object in the air only has one force acting on it: the gravitational force. This means the vertical motion has a constant acceleration and the horizontal motion has a constant velocity. The math required to solve this type of problem isn't very difficult. In fact, it's a pretty standard problem in just about every high school-level physics class.

With air resistance, there is now a force that depends upon the velocity of the object. Of course, the velocity depends upon the acceleration, so you get a loop of dependencies. The greater the speed, the greater the acceleration, which means the speed changes even more. It's a tough problem.

There is a solution. The method to solve this problem is to create a numerical calculation of the motion. An analytic solution (like the case with no air resistance) is solvable with some algebraic manipulations—or sometimes with calculus. The analytic solution is what you would typically see in an introductory physics textbook. For the numerical calculation, you need to break the problem into a bunch of small steps in time. For each step, you can assume the forces and the acceleration are constant. This means the typical, constant-acceleration solutions will work.

The smaller the time steps the problem is broken into, the better the solution. Of course, if you break a long jump into time steps that are one nanosecond in length, you are going to have to do one billion calculations for a one second jump. Even a time step of 0.01 seconds would require a hundred steps. Even this is too much for a person to reasonably do. The best bet is to use a computer, they rarely complain.

In order to see how much the changes in gravity and the density of air will affect a jumper, we need to start with a basic model. If we look at Beamon's record-setting jump, we can get some information about the initial velocity assuming there was no air resistance. From the video (and by counting frames), Beamon was aloft 0.93 seconds. Since he traveled 8.39 meters horizontally, this would put his horizontal velocity at 10.1 m/s (22.6 mph).

I can use a similar analysis with the vertical motion to determine that the initial vertical velocity for Beamon's jump was around 4.5 m/s. Now I can use these horizontal and vertical velocities and add the effects of air resistance and changes in the gravitational force.

If you make a plot of the trajectory for three cases—at sea level with no air resistance, at sea level with air resistance, and in Mexico City with both air resistance and slightly reduced gravity—you would notice something. There's not much of a difference, but there is a difference. The model with air resistance and at sea level gives a jump distance of 8.89 meters compared to Mexico City (with air) at 8.96 meters. That is just 7 cm further, but every little bit counts. In Beamon's case, it wouldn't have mattered if he made the jump at sea level or at 5,000 feet. He beat the previous record by an astounding 55 centimeters. That's truly an incredible feat. Bob Beamon was a true Olympic champion.

LINEAR REGRESSION AND THE LONG JUMP

There is another way to look at Bob Beamon's record-breaking long jump in the 1968 Summer Olympics. Suppose I create a plot that shows the date and jump distance every time the world record was broken.

Here is that plot for both men's and women's long jumps:

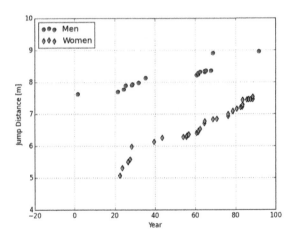

It always amazes me that there is a nearly linear progression of world records. Let me start with the women's records. It will be useful to find a function that fits this data. We call this process linear regression.

If I fit a linear equation to just the women's records, I get the following function:

$$s_w(t) = (0.0314 \text{ m/year})t + 4.656 \text{ m}$$

This linear model seems to fit quite nicely with the data. If you use the year (1967 would be 67 and 2012 would be 112), then the model will give you a predicted long jump record. What about the "4.656 m" in the equation? That is the modeled record at the year 1900. Of course, there weren't any records from then and I suspect they could jump farther than that.

Here is a fun thing: If I use this model and extrapolate all the way back to the time when the long jump record was 0.0 meters, that year would be 1885. Yes, that is silly. That's why this is just a model.

One other point: I can get a measure of just how linear this data fits the model with the correlation coefficient. This data gives a value of 0.98. A correlation coefficient of 1.0 would be a perfect fit. This means the data fits fairly well to a linear model.

Now for the men's records. Suppose I fit a function to everything but the last two records. Notice that Bob Beamon's record-breaking jump is the second to last in that plot. His jump was succeeded by Mike Powell in 1991. If I leave off these last two data points, I can temporarily avoid Beamon's unusual jump.

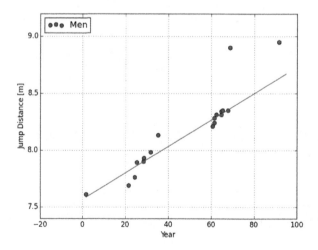

You can see that this linear function fits quite nicely if I do not include the last two record setting times. With this, I would get a function that has a slope of 0.0116 meters per year with an intercept of 7.57 meters.

It seems both Beamon's and Powell's records are "out of line." If all the records fit the above model, a long jump distance of 8.95 meters wouldn't be achieved until 2018.

Although these models mostly work, sometimes a new technique comes along to change the pattern. One example is the famous Fosbury Flop as used in the high jump. In the high jump, the athlete attempts to jump vertically over a horizontal bar. Before 1965, humans just did a normal jump and it seemed to work fine. However, in 1965 Dick Fosbury used a new technique. Instead of going over the bar facing forward and feet first, he turned his body so he went over facing backwards and head first. This new technique allowed him to break the world record and throw off the previous record-setting trends.

I am not sure if Beamon and Powell used a different technique to set their records, but they are in a league of their own. Let's wait until 2018 to see if the old model still works, as that's about the time someone should match or break Powell's record.

Finally, let's look at the slope for the men's record (0.0116 meters per year) and the women's record (0.0314 meters per year). That's a pretty big difference. The women are increasing their record at a much faster pace than the men. If both of these models still hold up, how long will it be until the women are jumping as far as the men?

All I need to do is set up the jump distance for men equal to that for women and solve for the year. It's a simple two equations with two variables algebra problem. I will save you the trouble of solving this and just tell you the result.

If you put a time value of 147 years into both men's and women's records, they will both be at a record of 9.27 meters for the long jump. This would be in the year 2047 since I let the value of $t = 0$ be the year 1900.

Of course, I doubt these models will work so far into the future. We already know that in the year 2029 the Earth will be overrun with robots as foretold in the movie *Terminator*. We may not even have track and field events at that time. Or maybe we will decide to let robots compete in the Olympics. That would be a whole new set of data.

SCORING THE DECATHLON

In the decathlon, athletes compete in ten events (thus the "deca" part of decathlon). However, this presents a problem. How do you compare the performance of several individuals across ten different events? Even more problematic is that four of these events have results measured in seconds (for races) and six events are recorded in meters. Even for the distance-based events, it is difficult to compare the distance of the long jump to the range of a javelin.

What can you do to compare all these scores? The answer is to come up with some type of formula that takes the results from each event and returns a value. For a fair formula, equal weight would be placed in all ten events. This is a task easier said than done. However, there is a formula to calculate the score for each event. It looks like this[14]:

$$\text{Time Event Score} = S_T = A(B - P)^C$$
$$\text{Distance Event Score} = S_D = A(P - B)^C$$

The P is the result from the event and the A, B, and C are constants that change for each event. Notice there are two different formulas. For the time-based events, a lower time is better and leads to a higher event score since it is some constant minus that time. The opposite is true for distance-based events.

What about the units? Yes, at first glance that does look like a problem. First, it doesn't make much sense to have both B and P in units of seconds or meters even though you could subtract them in that case (you can't subtract a unitless quantity from one with units). If you did that, then the constant A would have ridiculous units like meters to the -1.81 power. Instead, I would suggest putting P/K where K is something like 1 meter. That way you remove units from the equation right away.

There is a very funny video about the scores in the decathlon. Lieven Scheire[15] shows that for very poor performances, you can end up with an imaginary score, as in the square root of a negative number. Actually, you will get a complex number (one that is part imaginary and part real).

---

14 From Wikipedia's Decathlon page http://en.wikipedia.org/wiki/Decathlon#Points_system

15 You can see the video on YouTube http://youtu.be/LHuNjHojurU

Unfortunately, I am pretty sure the official rules take this into account by saying you can't win this way. It would be really funny if someone actually did.

So which event is most important? Like I said, the goal is to make all events equal. But what if you break the world record in the 100-meter dash by 0.01 seconds? How would that help your score? What if instead you broke the 400-meter record by 0.01 seconds? Would that be better or worse for your overall score?

First, let's pick an event: the high jump. How would different distances produce different scores? The world record for the high jump is 245 cm. This plot shows the scores for heights from 50 percent of the world record jump to 105 percent of this height.

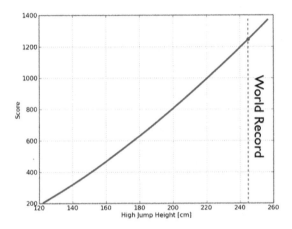

For this world record performance, you would get 1,244 points. Not too bad since the world record for overall decathlon score is just over 9,000 points. The high jump is a nice one to look at since its C coefficient isn't as close to a value of one as something like the shot put event's coefficient.

How can you compare distributions of scores for all the events? One way would be to reduce the performance score to a fraction of the world record. So, the world record 100-meter time of 9.58 seconds would have a unitless performance value of 1.0. If you broke the world record with a time of 9.59 seconds, this would be a performance value of 1.001. If you do the same thing with all the events, a comparison can be made for all the scores.

Here is a plot showing the score dependence on performance. Notice that for time-based events, a lower time is a higher score. That is why these lines have a negative slope.

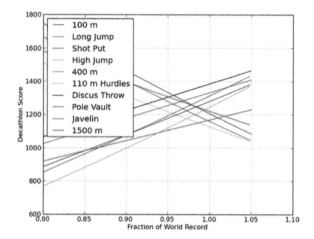

What can you learn from this plot? First, if you were going to tie the world record in one of these events it would be best to do it in the discus. A world record would give you a 1,382. Compare this to a tie in the world record for the 110-meter hurdles. That would only give you 1,123 points.

What about improving your performance? If you could increase your distance or time for one event, which should it be? Essentially this is looking at how much the score changes due to changing performance, which would be the derivative of each value. Taking the derivative of both formulas (with respect to P) would give:

$$\frac{dS_T}{dP} = -AC(B - P)^{(C-1)}$$
$$\frac{dS_D}{dP} = AC(P - B)^{(C-1)}$$

Of course, you could put in the values for A, B, and C to determine the rate these scores change. The rates aren't constant, but instead depend on the performance, so you would have to pick a performance value to plug in, like your own time or the world record time. However, there is a simpler approximation. Which event score plot above has the greatest slope around the world record score? For the time-based events this would be the 100-meter dash and for distance-based it would be the long jump. But remember, this graph shows the change in score for a corresponding change in performance based on the fraction of the world record performance. Is it as easy to do 5 percent better in the 100-meter dash as it is to do 5 percent better in the 1,500-meter run? No, it probably isn't. This is exactly why it is so difficult to get a high score for any event in the decathlon.

WHY IS IT HARDER TO SWIM FASTER?

It's hard to be a champion and harder still to set a record. Let me look at one event in particular, the 50-meter freestyle in swimming.

In this event, swimmers travel one length of the pool. They simply dive in and swim. There are no turns. Well, unless they're swimming a 50-meter short course, which uses a 25-meter pool. The cool thing about this difference is the men's record for the long course is 20.91 seconds and the record for the short course is 20.30 seconds. Clearly swimmers get a pretty good boost pushing off the wall in a flip-turn.

Using the world record time on the men's long course set by Cesar Cielo of Brazil in 2009, I can get an average speed for the race. He swam those 50 meters in just 20.91 seconds. This puts his average speed at 2.39 meters per second.

Perhaps you prefer different units. If so, this is about 5.3 mph. Of course, this includes the higher starting speed that comes with diving off the block. So it wouldn't be unreasonable to say an Olympic freestyle 50-meter swimmer could have a speed of 2.2 m/s in the water.

Still, everyone wants to go faster but what does that take? Let's think about the forces on a swimmer moving at a constant speed. In this model, the swimmer is moving at a constant speed which means the net force must be zero (technically, the zero vector). The forces in the vertical direction aren't important in this discussion, but let me say that the "up" force is a combination of buoyancy force and "lift" due to the motion of the swimmer.

Other forces include the drag force due to the collision of the swimmer with the water. This is why a swimmer does not continue accelerating throughout the race. It has some dependence on speed, but for now let's say it is in the opposite direction of the swimmer's motion. Additionally, the thrust force is a result of the swimmer using their arms and legs to propel their body through the water.

The important thing is that the thrust is a result of the swimmer's exertion, or using energy to move through the water. This is all about power. One way to think about power is to consider how much work is done in a certain amount of time. In this case, the work is the force exerted (the thrust) multiplied by the distance traveled. Yes, work is a little more complicated than that, but this definition will do fine here. (I was going to say this definition will work . . . get it?)

If you put the definition of work with the definition of power, you get a change in distance over time. So the power becomes the force times the speed. The power to swim in the race is independent of the actual distance.

Now, going back to the thrust, it has to have the same magnitude as the drag. But how would you model the force from the drag in the water? The drag force must, in some way, depend on the velocity of the swimmer. And how would you model this speed-dependent drag force? The best way would be to create an experimental method of measuring this force. For this situation, let me assume the drag force is linearly proportional to speed.

If the thrust depends on the speed and the power is the product of speed and force, then the power would be proportional to the square of the speed. So, if you want to go twice as fast, it doesn't take twice the power. It takes four times the power. Like I said, going faster is difficult.

So how about some values? First, what kind of power can a human produce in short bursts? This is a difficult question since it depends on how the person is moving. Also, power isn't such an easy thing to measure. The paper "Laboratory Measurement of Human Power Output during Maximum Intensity Exercise"[16] suggests a maximum power of around 1,200 watts for short periods. If I use this and Cielo's world record speed, I get a value for this drag coefficient (b). The drag coefficient is the term in front of the squared velocity for the power. It depends on the shape and size of the swimmer as well as the type of swimsuit worn. This gives a coefficient around 248 kilograms per second.

Now suppose you want to break Cielo's record and swim with an average speed of 2.21 m/s instead of 2.2 m/s. How much power would you need? Using the same coefficient of drag, you would need 1,210 watts instead of 1,200 watts. If you look at the increase in speed, it is just a 0.5 percent increase. If you look at the power it is a 0.8 percent increase. That may seem small from

16 http://iopscience.iop.org/0031-9120/28/6/007;jsessionid=50E7DB2CEEC-8594B07A311CE28D51D17.c2

the comfort of your couch but in the water, when you're pushing the limits of human performance, it is big. This is why it is so difficult to break a world record time, especially in swimming.

WHAT'S THE STEEPEST GRADIENT FOR A ROAD BIKE?

Why would cyclists get off a bike and push it? In the case of a recent stage of the Tirreno-Adriatico, there were three parts of the course with a 27 percent gradient which caused many of the athletes to get off their bikes and walk them up the incline. Yes, a 27 percent gradient is pretty steep for a bike. But this brings up the question: what's the steepest gradient you could ride up?

I think there are two reasons why a slope would be too steep. For all of these cases, I am going to assume that it is a prolonged slope. This means you can't just build up a large speed and zoom up the slope. If this was the case, you could go straight up a wall (which you can, for a short time).

First, let's look at the gradient limit due to human power.
A diagram might help.

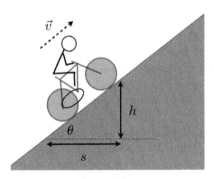

Before we start, a quick note on grades. What does a 30 percent grade mean? It means that if you travel a distance up the incline, the ratio of vertical to horizontal distance (times 100) would give you the grade. We usually represent the steepness of a slope with an angle, but this essentially does the same thing. I'm not sure of the international symbol for grade, so I will use r. In terms of the height and horizontal distance, the grade would just be 100 times the height divided by the horizontal distance.

Let's say that the rider is moving with some speed, v, and this speed is slow enough that air resistance isn't a significant factor. How much energy would it take to move up the slope at a constant speed? In this case, I could consider just the energy going into the change in gravitational potential energy of the rider plus the bike. The change in energy for this would be the product of three things: the mass of the bike and the rider, the vertical change in height, and the gravitational field which we usually call g.

Of course, I don't really care about the change in energy, I care about the power. Power is defined as the rate of change of energy (change in energy over change in time). You know what else depends on a change in time? Yes, velocity depends on time too. So if you put these two things together, the power is equal to the vertical component of velocity times the weight (mg) of the bike. The faster you move up, the greater the power needed. A steeper road means that a larger component of the velocity goes into calculating the total power.

What about the power needed to just move the bike? Sure, there are other things that make a human exert effort on a bike. There is the internal friction from the gears and pedals. Additionally, there is a frictional force from the rolling tires as well as an air resistance force. But for this calculation, I am going to be looking at the steepest possible incline. In this case, the rider should be moving slowly enough that air resistance will be negligible. The other effects should also be small compared to the total power needed to climb this hill.

Now I just need some estimates in order to calculate the steepest incline. Suppose a cyclist and bike have a mass of 75 kg and move with an average speed of 2 m/s. If the grade of the incline was thirty, this would require a power of 422 watts. That's some serious power. I'm pretty sure I could produce 422 watts, but only for a very short time.

My brother likes to bike quite a bit. He said that over a long ride he can average 280 watts. What if we bump this average power up to 300 watts (you know, because of professional cyclists)? With this power, there are two things that matter: the speed and the incline. If the cyclist is only going at a speed of 1 m/s then the steepest grade would be around 45 percent. By increasing the speed up to 4 m/s, the steepest incline drops to about 10 percent.

If I were designing a cycling path, I wouldn't have any part go over a 20 percent grade incline. At over a 20 percent grade, the cyclists would probably be better off just walking up the incline. And remember, this is supposed to be a bike race not a walking race. Why would walking be easier for these steeper grades? When you walk, your progress does not depend on your speed. You can go slower (so lower power) and not fall over.

Now for another consideration: the center of mass. For a bike going up an incline, the center of mass has to be horizontally between the two support forces. In this case, the support forces are the contact forces between the two tires.

Here is a diagram of a biker going up a hill—of course, if you are riding up a steep hill, you will be leaning forward some:

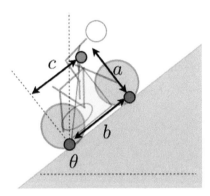

There are three important locations. First, there is a dot representing the center of mass of the bike plus the rider. Then there are two dots representing the contact points for the two tires. The most important thing is that the center of mass dot is in front of the back tire contact point.

So, how steep can this road be before the center of mass moves over the back tire? Let me make some assumptions. I will estimate the height of the center of mass at 0.8 meters and the distance from the center of mass to the back tire at 0.75 meters. This would mean that with a road incline of 43 degrees, the center of mass would be directly over the back tire. Any steeper and the rider would just flip over backwards. If you translate this slope into a gradient, that would be 93.7 percent gradient. Of course, I already calculated that the human power wouldn't let a human climb something this steep anyway.

CHAPTER 7: REAL SPACE

WHAT'S THE DIFFERENCE BETWEEN A SPACESHIP AND A SUBMARINE?

In the second *Star Trek* reboot movie, one scene shows a Federation space-ship going underwater. So, can a spaceship go underwater? You would think they could. I mean, what does a spaceship do? It is a vessel that allows humans to travel in an area where they couldn't otherwise survive. A subma-rine does a similar thing, right?

How do you keep humans alive? Well, we humans need a few basic things. We need food and water and the Internet, obviously. What else do we need? Oh, air. We need air. Technically, we need oxygen. However, there is a range of oxygen pressures that are bad for humans. Typically, if the partial pressure of oxygen is over 1.4 atmospheres, bad things can happen, like convulsions bad. On the other end, if the partial pressure of oxygen is less than about 0.16 atmospheres, people start to pass out.

What is the partial pressure of oxygen? Let's look at the oxygen in the air as an example and pretend that air is 79 percent nitrogen gas along with 21 percent oxygen (this is just an approximation). At one atmosphere of pressure, the par-tial pressure of normal air is 0.21 atmospheres. So, the partial pressure is the pressure of a gas if it was the only gas in the container. The partial pressure of nitrogen in this case would be 0.79 atmospheres.

Now, let's say I have a pressurized submarine that has air at a pressure of two atmospheres. In this case, the partial pressure of oxygen would be 0.42 bar (I am going to switch units of pressure to bar where one bar is approx-imately one atmosphere). But the question is, why would you increase the pressure inside a submarine? There is one other thing a human needs to survive. The human needs the walls of the capsule to not squash them. If you increase the pressure of air inside the submarine, the walls don't need to be quite as strong.

Suppose my submarine is in a cube shape and it is 1 meter by 1 meter by 1 meter. Yes, I know it would be a cramped space, but no one ever said subma-riners had an easy life. Now I take my cramped submarine down to a depth of about 10 meters. At this depth, the pressure pushing in on the walls of my awe-some vessel is about two bar. There is one bar of pressure due to the atmo-sphere above the water, and 10 meters of water adds about another bar. So, this two bar of pressure on a 1 meter squared surface gives a force of 200,000 newtons. Yes, that is a large force for the small wall of my submarine. However, there is some good news. If I have air inside the submarine, the air has a pressure of one bar pushing back out. This means that the net force on the wall would be 200,000 newtons pushing in minus 100,000 newtons pushing out for a net force of just 100,000 newtons.

Now, suppose I had very thin walls on my submarine. I could still go down to a depth of 10 meters if I increased the pressure inside the submarine. If the pressure inside was the same as the pressure outside, the net force on the walls would be zero newtons. Why not do this with all submarines? Well, there are two bad things with a design like this. First, what if you go down 60 meters deep? In that case, the pressure inside the submarine would have to be seven bar. Air at seven bar would produce a partial pressure of oxygen at around 1.4 bar. This is right on the limit of bad things happening to a human. If you want to go any deeper, you would have an even greater risk of oxygen problems.

I said there was another problem with having an increased pressure in a submarine, didn't I? The other problem is that your body absorbs the nitrogen you breathe in until the pressure of nitrogen in the body tissue is equal to the ambient pressure. However, this isn't the problem. The problem comes when you try to decrease the pressure. This results in a greater nitrogen pressure in your tissues so that the nitrogen "out-gases" into your blood. If this happens too fast, bad things happen. Scuba divers have to deal with these issues all the time. That's why they have a time limit on how long they can stay down and how long it takes them to return to the surface. If your submarine changes the interior pressure, you have to consider this. And, just so you know, there are submarines that do this. They are much cheaper to make but they don't go as deep.

What's the other option for a submarine? The other option is to build thicker and stronger walls for your vessel. If you build the walls strong enough, you can keep the interior pressure at one atmosphere. Of course, stronger walls mean more material and more mass. If you want to use this submarine as a spaceship, that might be a problem. The greater the mass would mean a much greater amount of fuel for a rocket to lift into space. When it comes to rockets, every kilogram counts.

If submarines don't make nice spaceships, what about the other way? Would a spaceship make a nice submarine? No, it wouldn't. Like I said before, spaceships need to have a low mass in order to get into orbit. Also, think about the role of the spaceship in terms of human survival. The ship needs to keep the pressure inside at about one atmosphere. Since there isn't any gas outside the spaceship, the same cubical design would have 100,000 newtons pushing out on the wall, not in. You would need to design a spaceship a little differently than a submarine to account for the direction of these air pressure forces.

Now, back to the Federation spaceship that goes underwater. Is there anything wrong with this? I am going to go with an answer of "no." First, from the short clip it doesn't seem like the spaceship goes very far underwater, so the pressure wouldn't be all that large. Second, it's a Federation spaceship. The thing has photon torpedoes and a warp drive. I'm sure it also has a strong hull. Who knows what that thing is made of? Or maybe as the spaceship goes underwater, the air pressure inside the vessel is increased. I understand how some people might object to this situation, but I think it could work.

CANDY IN SPACE

One of the European Space Agency's (ESA) spacecraft is the Automated Transfer Vehicle (ATV). The primary role of the ATV is to bring supplies to the International Space Station. Supplies include food, water, oxygen, scientific equipment, and candy bars. Yes, I listed food twice. Candy is food but I listed it separately so we could look at candy in space.

Think about this: suppose an astronaut requests an extra candy bar to be sent up on the ATV. This extra mass of the candy bar means there will be more energy needed to get into orbit. But how much more energy will you need?

To start, let's get some initial values. Some of these will be wild estimates but perhaps that is to be expected. Suppose the International Space Station (ISS) has an orbit that is 420 kilometers above the surface of the Earth and it moves at a speed of 7,700 m/s. Both the altitude and speed will be important in the calculation of energy needed to supply the ISS.

There is another important piece of information we need to have: the location of the ATV launch pad. It is located in Kourou, French Guiana. Kourou is only 5 degrees above the equator. There is a reason for this as we will soon see.

Oh, and one last thing. What about the mass of a candy bar? I don't wish to single out any particular brand of candy bar, so I will assume an average chocolate candy bar. Let's say it has a mass of 50 grams with 250 calories (that's food calories, which are different than chemistry calories, just to be clear).

What is the difference between food calories and science calories? They both are measurements of energy, but one food calorie is equal to 1,000 actual calories. I have no idea why the same word exists for the same thing. I suspect it has something to do with food and people that eat food. The standard definition of the calorie (sometimes called the chemistry calorie) is the energy needed to increase the temperature of 1 gram of water by 1 degree Celsius. That might be fine for chemistry, but in physics we prefer the joule as the unit of energy. One chemistry calorie is equal to 4.187 joules.

Now, for a little bit of physics. Why does it even take energy to get anything to the ISS? Well, there are two things you need to do to a candy bar in order for an astronaut in space to eat it. First, you have to lift the candy bar up to the height of the ISS. Second, you have to increase the speed of the candy bar so that it is going at the same speed as the ISS. Let me look at these two things separately.

Suppose you find this 50-gram candy bar on the ground and you lift it about 1 meter up to place it on a table. This requires that you do some work on the candy bar to change its energy. But how much energy would it take? One way to look at this is via the change in gravitational potential energy. On the surface of the Earth, the change in gravitational potential can be calculated as:

$$\Delta U_g = mg\Delta y$$

Here, g is the local gravitational constant with a value of 9.8 newtons/kg. Increasing the height of a candy bar by one meter would take 0.49 joules of energy. That's not too much.

What if I want to increase the height of the candy bar all the way up to the ISS? Can I just do the same calculation but change the height from one meter to 420 km? No, I can't. The above model for gravitational potential energy assumes the gravitational force on the object is constant. This is a good assumption near the surface of the Earth, but not so good as you get higher (though at the ISS's height, it isn't the worst approximation you could ever make).

If we use a better model for the change in gravitational potential, it would be this:

$$\Delta U_g = -G\frac{mM_E}{R_E+h_{ISS}} + G\frac{mM_E}{R_E}$$

Here, *G* is the universal gravitational constant. The two masses in the expression are the mass of the candy bar and the mass of the Earth. The values on the bottom of the expression are the distances from the center of the Earth. So, the candy bar ends at the altitude of the ISS (I call this *h*) and starts at the radius of the Earth.

If you put in the values for *G* and the radius and mass of the Earth, you would find that it takes 1.93 x 105 joules of energy to get the candy bar up to the right altitude.

However, that's not all of the energy for the candy bar. If you put that much energy into the candy bar and let it go, it would just fall from space back to the Earth. The other kind of energy the candy bar needs is kinetic, i.e. moving, energy. The kinetic energy of an object is just a factor of one-half multiplied by the product of the mass and the square of the velocity.

Since we know the speed of the ISS, shouldn't this be easy to calculate? If I put in the mass of the candy bar (0.05 kg) and the speed of 7,700 m/s, I get a kinetic energy of 1.48 million joules. Actually, this is the wrong answer. Why? It assumes we took that candy bar and increased its speed starting from rest. The only problem is that before the launch, the candy bar is already moving. It is moving because it is on the rotating Earth.

Let's say the Earth rotates once every twenty-four hours (which it doesn't actually—that's the time it takes the Sun to return to the same position—but this value is close enough for us). This means the speed of the candy bar before launch is just the circumference of the Earth at that latitude divided by twenty-four hours.

Why does latitude matter? Think about the radius of this circle the candy bar moves in. At the equator, the circle radius is the same as the radius of the Earth. However, at the North Pole, the candy bar wouldn't be moving in a circle at all. It would just be spinning around in place. Hopefully at the North Pole Santa Claus wouldn't eat the candy bar. You know how much he loves sweets.

Because Kourou is very close to the Earth's equator, the circular radius is basically the same as the radius of the Earth. This gives an initial speed of 464 m/s. That value might be small compared to the ISS' speed, but every little bit helps. This is why the ESA launches the ATV from Kourou instead of somewhere in Europe.

Okay, so what about the new change in kinetic energy of the candy bar? Launching from the equator, you would need about 1.47 million joules.

The total energy needed to get this candy bar to the ISS is the sum of the two values we've now calculated: the change in kinetic energy and the change in gravitational potential energy. This works out to 1.66 million joules. That is over a million joules of energy for just one tiny candy bar and it assumes a perfectly efficient method for getting things into orbit without any energy loss. This is why we don't all live in space. It's just expensive.

It's pretty hard to get a feeling for an energy amount of one million joules. What about a comparison to the energy in the candy bar? If you consume the candy bar it can produce 250 food calories. One food calorie is 1,000 calories which is 4,180 joules.

Let's go backwards. If it takes 1.66 million joules to get a candy bar into orbit, how many food calories is that? This is a pretty straightforward unit conversion problem. Remember that the trick to unit conversions is to always multiply by a fraction that is equivalent to one. For example, I can convert 1.2 feet by multiplying the fraction 0.3048 meters/one foot. Since one foot is the same length as 0.3408 meters, this fraction is equivalent to one and doesn't change the actual answer and instead changes the units.

With this unit conversion, I get an energy value of 1.6 candy bars. Well, that's not so bad. It takes a little more than one candy bar of energy to get an actual candy bar to the ISS!

One last question: What if we wanted to get all of the ATV cargo to the ISS just using candy bars as energy? The ATV with a full payload has a mass of about 20 tons, or 20,000 kilograms. If this payload was comprised just of candy bars, that would be 400,000 bars of candy. The energy needed to get into orbit is the same energy you would get from consuming 640,000 candy bars.

Recently, the United States government declined an online petition to build an actual Death Star[17]. Yes, it's a shocking decision in many ways. But let's pretend for a moment that someone did want to build a Death Star. Could you use the Automated Transfer Vehicle to do this? Well, of course you could, but what would that be like?

If you want to estimate how many ATV launches it would take to build or supply the Death Star, you first need to know something about the Death Star. I could make some wild estimates about the Death Star, but I won't. Instead I will look at two interesting estimations. The first is from an economics blog, Centives[18]. Centives starts their estimate in the same way I would by assuming the Death Star is kind of like an aircraft carrier. The idea makes sense. With this assumption, they get a mass of 10^{18} kg of material needed to construct the Death Star with a diameter of 140 km. Just a quick point (or reminder): There were two Death Stars. According to Wookieepedia[19], the first Death Star had a diameter of 160 km while the second Death Star was even bigger. I will consider building the smaller Death Star (but I would *not* put the exhaust port in a position that is so easy to send a proton torpedo down, that would just be foolish).

Okay, but there is another estimate of the mass of the Death Star. The popular site io9[20] assumes the Death Star wouldn't even be built of something like steel. Modern warships are built of steel since they have to be rigid enough to withstand waves and torpedoes. The Death Star would be in space, so it wouldn't need to be as rigid to prevent collapse due to water pressure. Also, the Death Star would likely have some type of force field as a shield to protect from enemy fire. Perhaps the Death Star would use steel, but only on the exterior surface. The interior could be constructed from a much lower density material like something similar to carbon fiber.

Let's go with the following estimate for the mass of the Death Star: if it has a 10 cm thick outer shell of steel, the mass would be about 6×10^{13} kg.

---

17   http://www.popsci.com/science/article/2013-01/white-house-shoots-down-death-star-petition-and-its-awesome

18   http://www.centives.net/S/2012/how-much-would-it-cost-to-build-the-death-star/

19   http://starwars.wikia.com/wiki/Death_Star

20   http://io9.com/5979110/how-much-would-a-death-star-really-cost

Now, what about the interior? Let's say the rest of the interior of the Death Star is somewhat similar to the density of the International Space Station. That makes sense, right? The ISS has a pressurized volume of 837 m³ and a mass of 4.5 x 10⁵ kg. This would give an approximate density of about 550 kg/m³. Of course, this is wrong since there is more mass than just the inside pressurized part of the ISS, but at least it's a start. If I use this same density for the interior of the Death Star, I get a mass of 10¹⁸ kg. This more or less agrees with Centives' estimation (the mass of the steel is negligible compared to the interior).

Now what about the ATV? The current version of the ATV can carry a payload of 7,200 kg. How many ATV launches would be required to bring all the material needed into orbit (assuming the Death Star would be built in the same orbit as the ISS)?

$$N = \frac{m_{\text{Death Star}}}{m_{\text{ATV payload}}} = \frac{10^{18} \text{ kg}}{7.2 \times 10^3 \text{ kg}} = 1.39 \times 10^{14}$$

That is clearly more ATV launches than anyone would like to deal with. If one ATV mission was launched every month (very ambitious), that would take 10¹³ years. Just for comparison, the Earth is under five billion years old. Our Sun will only last maybe another five billion years. 10¹³ years is just a little too long for this Death Star to be built. By the time it was finished, our star would be dead. Get it? Death Star? Yes, that was a bad joke.

Let's look at this from the other end. Suppose we want the Death Star finished in 10 years. If we still used the ATV, we would still need the same number of launches. How frequent would these launches be if they were evenly spaced out? This would be 1.38 x 10¹³ launches per year. It's difficult to picture that many in a year. If I convert this number, I get four launches per second. Clearly, the current ATV infrastructure can't handle this turnaround time. What if the whole human race works together to ramp up the ATV process? Maybe each ATV can relaunch with a two-week turnaround. During this two-week period, there would need to be 4.8 million different ATVs. That's quite a large number of vehicles.

The primary role of the ATV is to bring supplies to the ISS. What if the Death Star was already in orbit, how many ATVs would be needed to service it? Prepare yourself for some more wild estimates. First, I need to get a value for the number of crew members on the Death Star. What if the Death Star has a similar crew-volume-density as a *Nimitz*-class aircraft carrier? Why would that be the case? Why not? According to this site[21], a *Nimitz*-class carrier has a crew of over 6,000 people with a volume of about 2 million cubic meters. This gives a crew density of around 0.003 crew per cubic meter. With this same density, the Death Star would have a complement of 6×10^{12} people. That is crazy since there are only seven billion people on the Earth.

We are just going to have to be a bit reasonable with this crew estimate. There could possibly be large portions of the Death Star that just aren't used at the same crew density as an aircraft carrier. I know this feels like cheating, but I am just going to say there are one million people on the Death Star.

So, how many ATV missions would you need for a space station with one million people? The ISS gets about one ATV supply every six months for a full crew of six astronauts. That would mean about one ATV per month, per astronaut. If I use this same ATV supply rate for the Death Star, it would be one million ATV launches per month.

Of course, a station the size of a small moon might be able to produce many of its own resources. There is probably a greenhouse, a bakery, and probably even a candlestick maker. Maybe they just need mail and shipments from their online purchases delivered via the ATV.

---

21   http://www.naval-technology.com/projects/nimitz

CHAPTER 8: CRAZY ESTIMATIONS

CAN ICE CREAM GET COLD ENOUGH TO BE ZERO CALORIES?

There was a very funny series of physics-inspired cartoon diagrams by Christoph Niemann at the *New York Times*. I would like to look at one of these in a little more detail.

The basic idea of the cartoon is that if you make ice cream cold enough, your body will do two things. First, it will metabolize the ice cream and "gain" calories. Second, it will warm up the ice cream to body temperature and thus "use" calories. If these two have the same value, the net change in calories is zero. Simple, right? Niemann says the ice cream would have to be at a temperature of -3,706 °F for this to work (which is not possible). How about I redo the calculation just to see what happens?

To start, we need to look at temperature and thermal energy. What is temperature? You might be surprised that this isn't as easy to define as you would think. Often it is described as a measure of the average kinetic energy of the particles that make up an object. This isn't a terrible description, but it isn't my favorite. I like to say that temperature is the physical quantity that two objects have in common when they are in contact for quite some time. It's true. If you take a small wooden block and a large metal block and place them together, they will eventually reach the same temperature.

What about thermal energy? This is a measure of exactly how much energy an object has due to its temperature. Consider a small wood block at 70 °F and a much larger metal block at 10 °F. Although the wood block has a higher temperature, it is possible that the metal block actually has more thermal energy. The thermal energy in an object depends on its temperature, its mass, and the type of material. The energy dependence on the type of material is called specific heat capacity.

There are two problems with this model for the change in thermal energy. First, it isn't valid to assume the specific heat capacity is constant over a large change in temperatures. Also, it does not include the energy needed to go through a change of phase (such as from solid to liquid). I can easily add in the phase change energy, but I will pretend like the specific heat is constant.

So, suppose you consume some ice cream that has a mass m and metabolizes E_F energy from the food (the "F" stands for food). If you want a net change in energy of zero, then this energy must go into two things: heating up the ice cream and melting it.

All I need to do is solve for the temperature such that the energy is the same as the food energy you get by consuming the ice cream. That is it. I just need to estimate some values. First, let me assume that ice cream is like water but with flavor and calories. This means that the specific heat would be one calorie/gram and the latent heat of fusion would be about eighty calories/gram. By the way, I hate the unit of calories for energy. To me, this is like using the unit of slugs for mass. However, in this case it is the usual unit of energy for food. Except, one food calorie equals 1,000 calories. I don't know what genius came up with that idea.

Now for some ice cream. How much food energy do you get from eating this stuff? Here is one of many online calorie calculators[22] that lists 72 grams of vanilla ice cream as having 145 food calories (1.45×10^5 chemistry calories). Using that and a final temperature of 37 °C, I get an initial temperature of -1,900 °C. Of course, that temperature is impossible since -273 °C is the lowest possible temperature.

So, two things: First, how does this compare to the value from Niemann? Second, how could I make this work?

Niemann gave an initial ice cream temperature of -3,706 °F which is around -2,000 °C. He probably didn't take the phase change into account, so if I remove my phase change term, I get just about -2,000 °C.

How could I make it work, though? Well, I know how much energy I can put into increasing the temperature of the ice cream. I just need to make it the same as the food energy. Let the initial temperature be -273 °C, and then the ice cream can have 2.8×10^4 calories, or 28 food calories. Looking at the other ice creams on thecaloriecounter.com, not even the light, sugar-free ice cream is around twenty-eight calories. I guess we will have to wait for science to catch up to this awesome idea.

---

22 http://thecaloriecounter.com

HOW MANY DOLLAR BILLS WOULD IT TAKE TO STACK THEM TO THE MOON?

I was watching a political discussion show on TV. Yes, it happens every once in a while. In this particular show, they were discussing the national debt and funding different programs. One participant claimed that if you stacked one trillion dollars (in one dollar bills, I assume) it would go to the Moon and back four times.

It's not that I don't trust what people say on TV, but this seems like something that could be estimated.

How thick is one dollar? I don't usually carry cash in my wallet, but when I do I measure it. There were five bills. I measured the thickness of just one, then two, and so forth. After plotting the thickness of the stack of bills versus the number of bills, I get a straight line. The slope of this line is 0.1 millimeters per bill. So, that would be a great estimate for the thickness of just one dollar.

Now, what about the stack of one trillion dollars? First, what is one trillion of anything? Sadly, not everyone agrees. In the United States, one trillion is 1,000 billion or 10^{12}. In some countries, one trillion means one billion billion, or 10^{18}. Confusing, I know. For this calculation, I will assume the 10^{12} version of trillion (since the show was in the United States).

So, if I stack 10^{12} bills, how high would it be? First, let me assume the bills don't get compressed. Why am I assuming that? I don't know, but you have to start somewhere. The height of this stack would be the thickness of one bill times one trillion. This gives a height of 100 million meters.

The distance from the Earth to the Moon is about 400 million meters. Okay, now there is a problem. According to my calculations, the stack of one trillion dollar bills would go one fourth of the way to the Moon. On the show, they said it would go there and back four times (which would be 32×10^8 meters).

Let me try one other thing. If one trillion dollars goes to the Moon and back four times, how thick would it have to be? All I need to do is take four times the distance to the Moon and divide by one trillion bills. This gives a bill thickness of 3.2 millimeters.

A dollar that was 3 millimeters thick would be rather awkward. So, I think they messed up their trillion dollar statement. It happens to the best of us.

Okay, so they were wrong. What else could I do with a stack of one trillion dollars? I wonder if it would even be possible to stack something this high.

Suppose you could stack the bills perfectly. As the stack gets higher and higher, it would be more likely to fall over from a slight nudge. Let's look at a diagram of stacks of bills:

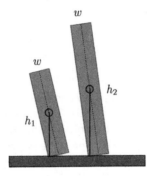

For each stack, the dot represents the center of mass. If the stack is tilted such that the center of mass goes over the edge of the base, the stack falls over. Yes, I am assuming the bills stick together. But you can see that the taller the stack gets, the smaller the "tilt" angle would be for it to fall over.

Calculating this "tipping angle" for different stacks, I find that a 10-meter tall stack of bills would only need to be tilted 0.37 degrees for it to be at the tipping point. What if I go even higher (you know we couldn't get much higher)? Here is a log plot for stacks up to 10,000 meters:

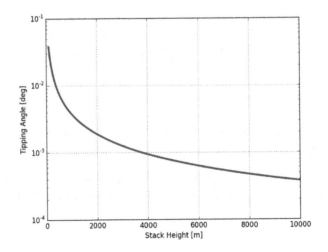

Ok, what if I take this up to 10^6 meters tall? That would be a tipping angle of 3.8 x 10^{-60}. And a trillion dollar stack (assuming it was all in a constant gravitational field, which it wouldn't be) would have a tipping angle of 3.8 x 10^{-80}. This tiny angle corresponds to a horizontal nudge of just 6.6 cm.

Even if the stack didn't tip over, is this stacking even possible? Would the bills at the bottom be able to support the weight of all the bills above it? This has to do with what we call "compressive strength." Essentially, the paper can only take so much pressure before something bad happens.

I don't know about paper, but wood has a compressive strength of 3 to 37 megapascals. Let me just randomly choose 20 MPa as the compressive strength of a dollar bill.

What is the pressure at the bottom of the stack? Well, it would be the weight of everything above it divided by the area of the bill (about 6.6 cm by 15.6 cm). This means the pressure would increase linearly with height (assume a constant gravitational field, which isn't quite true).

Using this pressure and an estimated bill density of 958 kg/m³, the pressure at the bottom of a stack of one trillion dollars would be 970,000 MPa. However, the pressure would be smaller than this because the gravitational field gets weaker as the stack gets higher. I don't think it matters. This pressure is way over the 20 MPa I estimated for the compressive strength.

Well, if stacking won't work, I am going to make a trillion dollar asteroid. I know the density of a dollar, so I know the mass of one trillion dollars. Why would you make a big ball of cash? Why wouldn't you? You could call it a cashteroid. Ok, first let me calculate the mass. If each bill is 6.91 x 10^{-3} kg, then 10^{12} of them would have mass of 6.91 x 10^9 kg. Assuming a constant density, this would give a volume of 7.2 x 10^6 m³.

If this is a spherical cashteroid, it would have a radius of 120 meters. This may seem like a small bundle of cash but that is 780 feet across. That might be hard to picture in your head, so maybe another comparison. This is about the size of four *Nimitz*-class aircraft carriers put together.

Maybe the people on the TV show should have paraphrased Dorothy Parker and said if the government took all the money it had and laid it end to end, I wouldn't be at all surprised.

The best questions come from the Internet. This was from a website called The Last Word[23]:

> *"How high would you have to drop a frozen turkey so that it is cooked when it lands?"*

Great questions like this require some starting assumptions. Here we go:

- The turkey is made of water and starts off as ice at 0 °C.

- The turkey is a sphere with a radius of 15 cm. (Or generically, I will call it r.)

- When the turkey falls through the atmosphere, half of the energy dissipated goes into the turkey and half goes into the atmosphere.

- I am going to ignore the increase in thermal energy when the turkey hits the ground. We just want to know if the turkey is cooked before it hits the ground and becomes turkey shrapnel.

- A turkey is ready if it is at 180 °F (82 °C).

Why do I even have to start with assumptions? Well, this is the one way to make a complicated question less complicated. It seems like cheating, but it isn't. Suppose I find that you have to drop a turkey from 100 meters to cook it (with my assumptions). This would suggest that the answer for a real turkey would probably be anywhere from 50 meters to 200 meters, but not 100,000 meters. That kind of answer (even though not exact) is still useful. And yes, I just made those numbers up.

Now that we have some assumptions, how about a diagram to go with it? Here is a spherical turkey dropped from a high altitude:

---

23 http://www.last-word.com

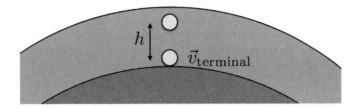

Why would this falling turkey even get cooked? Air resistance, that is why. As the turkey falls, the air hits it and produces a friction-like effect. Actually, the turkey heats from a combination of the air hitting it as well as the air in front of the turkey heating up as it is compressed. Yes, heating by air friction is a complicated thing. But the point is that moving through the air will make something hotter. Just as your hands get warmer when you rub them together, so would the falling turkey.

How would you even calculate this increase in temperature? When dealing with an object moving over some distance, the best thing to use is the work-energy principle. This says the work done on a system is equal to its change in energy. If the system is considered to be the Earth plus the turkey plus the air, then the following types of energy could be considered:

- Kinetic energy: due to the motion of the objects (mostly just the moving turkey).

- Thermal energy: both the turkey and the air warm up as the turkey falls.

- Gravitational potential energy: the potential energy will decrease as the Earth and turkey move closer to each other.

So, what does work on the system? In this system, only the air resistance force does work on the falling turkey. The typical model for the air resistance force has a magnitude that depends on the density of the air, the shape and size of the object, and the speed of the object.

You can test some of these air resistance properties yourself by just sticking your hand out of the window of a moving car. Clearly, the air pushes on your hand. As you go faster (increase v), the force increases. If you change your hand to make a fist, your cross-sectional area decreases (and the shape changes) and you will feel less force.

For the falling turkey, this air resistance force will increase as the turkey falls faster and faster. However, at some point the air resistance force and the gravitational force pulling it down will have equal magnitudes. This will make the net force on the turkey zero and lead to a constant velocity (also called terminal velocity). Here is a diagram showing this constant motion turkey:

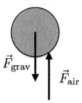

So what would be the terminal velocity for such a falling turkey? Well, if it is a spherical turkey the gravitational force would be proportional to the mass. If the turkey has a constant density (another assumption), this would also be proportional to the cube of the radius. The air resistance would be proportional to the cross-sectional area of this sphere (which is a circle). So the air resistance would be proportional to the square of the radius. The net result is that the size of the turkey matters (since the two effects of the size don't cancel). Bigger turkeys have a faster terminal velocity.

The dependence on scale is one thing that always surprises me and I am not sure why. I guess we tend to have this model railroad picture of the world. If you shrink things down to a smaller scale, it will just be the same, except smaller. But this isn't true. In fact, size does matter.

Now, back to energy. What about the gravitational potential energy? Essentially, the higher something is, the more gravitational potential energy it has. If things stay close to the surface of the Earth, this potential increases linearly with height. For very high distances it is a little more complicated. Just to be safe, I will use the more complicated version of the gravitational potential energy.

There is one more piece to this puzzle: the connection between thermal energy and temperature. Is there a difference between these values? Yes. Look at this common example (which I used in a previous chapter mostly because I love pizza). Suppose you want to reheat some leftover pizza. If you put the pizza on aluminum foil and place it in the oven, eventually both the pizza and the foil will reach the same temperature (say 150 °F). When it is ready, you can easily grab the aluminum foil with your bare fingers. But don't touch the pizza because it will burn you. Essentially, both of these things have different amounts of thermal energy.

When dropping the turkey we want it to reach a certain temperature (say 180 °F). The bigger the turkey, the more thermal energy required to get it to this temperature. So, again, size matters.

Another thing to keep in mind about thermal energy: both the turkey and the air increase in thermal energy. Just for simplicity I will assume half goes to the turkey and half to the air. Why did I pick half? Would it be fair if the turkey got more than its fair share of energy? No, half is fair. Actually, it's just an approximation, so half seems as reasonable as anything else.

Putting it all together with a typical-sized turkey, I calculate a height of 142,000 meters in order to cook the turkey. Is this answer suitable? No, not really. Why not? Because this is fairly high. Just compare this to the orbital height of the International Space Station at 300 km.

Other than the ridiculous height, what other problems come up in this situation? First, you might think that one would need to take into account the varying air density, especially if the turkey started as high as was claimed. Then again, it might not matter. If you have very low air density at higher altitudes, the turkey would start out with a much higher terminal velocity. However, the air would do less work at a higher altitude but make up for it at a lower altitude when the turkey is going faster.

What about uneven heating? Would this be a problem? Certainly. If you dropped the turkey from over 100 km, it wouldn't take all that long to hit the ground—most likely under ten minutes. What happens when you try to cook a turkey in ten minutes? The outside gets burned.

So for the next family gathering at Thanksgiving, you might be better off cooking the turkey in a conventional oven. Trust me on this one.

COULD YOU BUILD A SCALE LEGO MODEL OF THE DEATH STAR?

I was inspired. Inspired by this awesome estimation of the cost to build a Death Star from some students at Lehigh University[24].

No one on the Earth is likely to ever build a Death Star to see if they're right. However, Lego does make a version of the Death Star, though it doesn't even try to pretend it is to scale. You can barely fit a handful of minifigs (those

---

24 http://www.centives.net/S/2012/how-much-would-it-cost-to-build-the-death-star/.

little Lego people with the bulbous yellow heads), much less the amount that appear to man it in *Star Wars: A New Hope*. Lego does make many *Star Wars* ships to scale, though, including an older version of the Millennium Falcon it doesn't sell anymore (the Ultimate version). But would it even be possible to build a scale version of the Death Star?

First, how big is the real Death Star? Well, there were two (in *Episode IV* and *Episode VI*). Apparently these two Death Stars were not the same size. According to Wookiepedia[25], the first Death Star had a diameter of 160 km. A clone trooper minifig is 38.6 mm tall, not including hair, a helmet, or the little piece at the top you'd connect those to. If I assume an average human height of 1.77 meters, this would mean the ratio of minifigs to humans is 0.022.

So, a Lego Death Star (first version) built to scale would be 0.022 times the diameter of the "real" Death Star. This would put the diameter of the Lego Death Star at 3.52 km, or a bit over two miles. That's a pretty big Lego model.

If the scale version of the Death Star came in a set, how many pieces would it have? This is a tough one to solve. The first question we need to answer is: what will be on the inside of the Lego Death Star? There has to be something inside the structure to support the outside. If you want a scale model of the Death Star, you probably want everything, garbage compactor and all.

Assuming the inside of the model has structure, I need to get an estimate for the density. Let's go back to the Ultimate Millennium Falcon model. According to the website brickset.com, the model has 5,195 pieces. It has dimensions of 84 cm x 56 cm x 21 cm. If I assume the shape is rectangular, I can determine the Lego piece-density by dividing the number of pieces by the volume: 52,400 pieces per cubic meter.

This is just an estimate, but one I am fairly happy with. Sure, there are some large pieces in the Millennium Falcon model but there are also some small ones. I guess it is possible the Death Star would have a lower piece-density if it had more large pieces.

Using this density and the volume of the proposed Ultimate Death Star model, I can get the number of pieces in the set. A Lego model that is spherical with a 1.76 km radius would require 1.2×10^{15} pieces (1,200 trillion). It is possible that the Ultimate Death Star set has more of the larger pieces (and thus a lower piece-density). Let me just use a lower estimate of one hundred trillion pieces in the set.

---

25 http://starwars.wikia.com/wiki/Death_Star

What about the mass of this scale model? For this, I need the mass-density and not the piece-density. Again, I can get an estimate for the mass-density by looking at another Lego set. The Ultimate Millennium Falcon is listed at a shipping weight of 24.2 pounds. This must also include the box and the instructions, so maybe the pieces would weigh around 21 pounds (9.5 kg). This would give a mass-density of 96.2 kg/m³. Just a quick check on the Lego Death Star II (from *Return of the Jedi*): it has a mass-density of about 85 kg/m³ and it isn't even complete (but it was fully functional).

Let me just go with a density of 90 kg/m³. With this density (mass-density) my Super-Ultimate Lego Death Star will have a mass of 2.1 trillion kilograms.

Well then, how much would this Ultimate Death Star cost? I am going to go with the data I have. Since I previously looked at the cost of a Lego set as a function of the number of pieces in the set, I know that set will cost about $0.098 per piece.

It might be a stretch, but if I assume the price per set is linear all the way up to huge sets, this would put the cost right around 10 trillion dollars (shipping not included).

There is something else to consider. Where would you put this scale model of the Death Star? Somewhere on the surface of the Earth would be a bad option. The biggest problem would be support. Suppose I built a base to hold up the Death Star that was 0.3 km across. It may be 3.5 kilometers at its widest point, but remember it is a sphere. All of the weight of the Death Star would have to be supported on top of this base. This would result in a compressive pressure of 240 megapascals.

Let me give a quick refresher on pressure. Suppose you place your hand on the floor and someone without shoes steps on your hand. It might just slightly hurt, right? Now suppose this same person puts on high heels and steps on your foot with just the pointy heel. Just imagine this, but don't do it. Don't do it because it would hurt. So, what is the difference? In both cases, the force on the hand is the same but they are applied over different areas. In the case of the high heel the area is quite small, resulting in a larger pressure. Now, what about pascals? A pascal is just one of the units of pressure, like pounds per square inch (psi). One pascal is equal to 1 newton per square meter.

What about maximum compressive strength? The maximum compressive strength is the largest pressure a material can withstand before breaking in some way. Suppose you push on a rock with a toothpick. The rock will probably be fine. Now push with the same force on a block of Jell-O, it will "break." So, what about the 240 megapascals from the Death Star model? This is greater than the maximum compressive strength of granite. The base would crack for most materials, not to mention the structural integrity of the Lego blocks at the bottom of the model.

The only reasonable option would be to put this thing in orbit around the Earth, perhaps low-Earth orbit at about 300 kilometers above the surface. For a 3.5 km diameter object 300 km away, it would have an angular size of 0.67 degrees. This is just a little bit larger than the angular size of the Moon.

Wouldn't that be cool? People would mistakenly think the Lego Death Star was the Moon, just like Han Solo did.

JUMPING OFF A BUILDING WITH BUBBLE WRAP

How much bubble wrap do you need to survive jumping out of the sixth floor of a building? Let me roughly say it would be a height of 20 meters.

Where would you start with a question like this? Well, first we need some bubble wrap. What properties can I even measure from bubble wrap?

First, I can measure the thickness of a sheet of bubble wrap. Yes, there are all different kinds of bubble wrap but I'm using mine (you have to start somewhere). Instead of just measuring the thickness of one sheet, let me make a plot of total thickness as I stack several sheets.

I measured the stack every time I added a layer and plotted a graph with height on one side and the number of sheets on the other. The slope of this linear fitting equation is 0.432 cm/sheet and this will be a good estimate for the thickness of one sheet.

Next, I need to see how "springy" the bubble wrap behaves. Is it like a spring? If so, how stiff is it? If it were an actual spring, I would just add some weight to it and see how much it compressed. Let me do exactly that.

If you plot this, it looks fairly linear—proportionally more force leads to proportionally more compression. So, I would have to say the bubble wrap does behave like a spring. I can model the force the bubble wrap pushes on other stuff (like a person) as though it were a spring. The force would be proportional to the amount the bubble wrap is compressed.

From this data, I have found an effective spring constant for bubble wrap of this size. But what about other sizes? Suppose I have two sheets of bubble wrap stacked on top of each other instead of just one. If a mass is placed on top, each sheet will be compressed the same amount as just one sheet, since they both have the same force pushing down. But two sheets compressing gives an overall larger compression than one.

What if I compare a small single sheet and a larger sheet of bubble wrap? This would be like two sheets of bubble wrap next to each other. When a mass is placed on top, they both push up on the mass so that each would only have half of the force compressing it. So, two sheets side by side wouldn't compress as much as one single sheet.

In short, the bigger the area of bubble wrap, the more the bubble wrap acts like a stiffer (higher spring constant) spring. The thicker the stack of bubble wrap, the smaller the effective spring constant.

The property of a material that shows its stiffness, independent of the actual dimensions of that material, is called Young's modulus. Since I know the size of my sheet, I get a Young's modulus of 4,319 N/m^2 for this particular bubble wrap.

What about the jumping part? It isn't the jumping that is dangerous, it's the landing. The best way to estimate the safety of a landing is to look at the acceleration. Fortunately, I don't need to collect experimental data on the maximum acceleration a body can take, NASA already did this. Here is essentially what they came up with (from the Wikipedia page on g-tolerance[26]):

Time (min)	+Gx ("eyeballs in")	-Gx ("eyeballs out")	+Gz (blood towards feet)	-Gz (blood towards head)
.01 (<1 sec)	35	28	18	8
.03 (2 sec)	28	22	14	7
.1	20	17	11	5
.3	15	12	9	4.5
1	11	9	7	3.3
3	9	8	6	2.5
10	6	5	4.5	2
30	4.5	4	3.5	1.8

26 http://en.wikipedia.org/wiki/G-force#Human_tolerance

From this, you can see a normal body can withstand the greatest accelerations in the "eyeballs in" position. This is the orientation such that the acceleration would "push" the eyeballs into the head. In the case of jumping, this
means landing on your back.

There is a slight problem. If a jumper is wrapped in bubble warp, the acceleration during the collision with the ground wouldn't be constant. Here is a diagram showing a person covered in bubble wrap while impacting the ground:

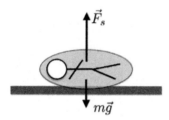

So, there are two main forces on the jumper during this time: the force from
the bubble wrap (which is like a spring) and the gravitational force. In order for
the jumper to stop, the acceleration must be in the upwards direction and the
bubble wrap force has to be greater than the gravitational force.

The acceleration depends on the value of the spring constant as well as
the distance the spring is compressed. I don't know either of those values.
However, if I use the work-energy principle then I can look at the entire fall. At
both the start and finish of this fall, the kinetic energy is zero. The gravitational
potential energy will decrease during the fall, and the spring potential energy
(in the bubble wrap) will increase during the impact. Since there is no other
work done on the system, I can create a relationship between the height of the
jump and the needed spring constant (for the acceptable acceleration).

In order to get a value for k, I need some other values.
Here are my assumptions:

○   Mass of the jumper plus bubble wrap is 70 kg. Here, I am assuming
      that the mass of the bubble wrap is small compared to the jumper.

○   Maximum acceleration of 300 m/s² and also an impact that lasts
      less than one second.

○   Starting height of 20 meters.

With this, I need a bubble wrap spring constant of 1.7 x 10⁴ n/m in order to have an acceleration during the landing that won't exceed the NASA recommendations for g-tolerance.

Now that I know the spring constant needed to stop the jumper, I am one step closer to determining how many layers of bubble wrap would be needed. There is one thing I need to estimate first: the area of contact between the ground and the bubble wrap. I know this area should actually change during the collision, so I am just going to estimate it. Suppose the contact makes a square about 0.75 meters on a side. This would give an area of 0.56 m².

Because I know Young's modulus for bubble wrap, I can calculate the thickness (which I have oddly called L) to be 0.142 meters. Since each sheet is 0.432 centimeters thick, I would need thirty-nine sheets.

Maybe thirty-nine sheets seems a bit low. Let me calculate the mass of this bubble wrap and how big it would look. If I assume the bubble wrap is wrapped cylindrically around the jumper, it would look something like this:

When looking down on a person, that person roughly appears as a cylinder with a radius of 0.3 meters (just a guess). If the bubble wrap cylinder extends another 0.142 meters, then what is the volume of bubble wrap? Oh, I guess I also need to have a person with a height of about 1.6 meters (another guess). This would give a bubble wrap volume of 0.53 m3.

I can use the thickness of bubble wrap along with mass data and find the density of bubble wrap. With this volume of bubble wrap, it would have a mass of 9 kg. Not too bad, but technically this would change the amount of bubble wrap needed to land. Maybe I could just be on the safe side and add a couple of layers to compensate for the added weight of the bubble wrap.

Now that I know the size of this bubble-wrapped person, I can consider the air resistance on the fall. The typical model for air resistance shows a force proportional to both the cross-sectional area and the square of the speed of the object.

Because the force changes with speed, the problem isn't so simple to solve on paper. However, it is pretty straightforward to calculate by breaking the problem into many short time intervals with a computer.

Doing this numerical calculation, I get the following plots for position and vertical speed of the falling object

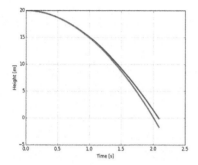 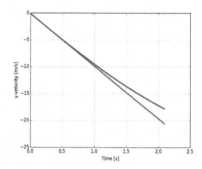

This shows that the falling object, with air resistance accounted for, ends up with a slightly lower speed before impact (17.8 m/s instead of about 20 m/s). I could redo all the calculations, but I won't. Instead, you could just consider this lower speed part of the safety margin (although I would never consider doing something like this "safe").

What about a follow up question (for your homework)? How much bubble wrap would you need to survive a jump out of a plane? I suspect it wouldn't be that much more. If you add some layers to the bubble wrap, you will decrease the terminal speed of this falling object.

Should we really use bubble wrap for human safety situations, though? Clearly, that would be a bad idea.

BANANA-POWERED GENERATOR

Bananas. You love them. You know you do. Really, it's not your fault. Who doesn't love bananas? It's even fun to say. Oh, I get it. You hate the taste. Fine, maybe you don't love *eating* bananas, but you still love them. Bananas are awesome. Here's why:

The first awesome banana fact doesn't have anything to do with physics. However, it is a fun thing to share at parties and while at the playground with your kids. Talking about the weather is boring, banana talk is much better. Here it is: Bananas are clones.

It turns out that wild bananas aren't very tasty, they aren't always very sweet, and they have huge seeds in them. But every once in a while, humans get lucky. We stumble on a mutant banana that has no seeds and tastes great. The problem is that you can't make more of these mutant bananas since there aren't any seeds. To fix this, humans just take a part of an existing mutant banana plant and replant it. Boom, you get more banana plants. Cloned, mutant banana plants.

Unfortunately, cloned banana plants mean that all bananas are genetically the same. If a mutant disease comes along that fits perfectly with the banana plant, you have what we call "a bad thing" and all the banana plants could be ruined. Could this actually happen? Yes, it's happened before. Have you heard the song "Yes! We Have No Bananas"? Guess why someone wrote that song.

That's enough about clone bananas. It's enough for you to start an interesting conversation, but not deep enough for me to say something that is just flat-out wrong.

If you thought being cloned and mutated made bananas cool, you are only partially correct. There is something even cooler. Bananas are radioactive. Calm down, you can still eat them. There are lots of things that are radioactive. It doesn't mean they are on the level of nuclear waste, but they are radioactive. Bananas are radioactive because they contain potassium.

What does the term radioactive actually mean? It means there is some type of nuclear reaction going on in the material. In a nuclear reaction, the nucleus of an element changes into some other element. If the element is a different mass (and they usually are) then, at the very least, energy will be produced and most likely some other types of particles.

Why is potassium radioactive? There are actually three different types of potassium. All potassium has nineteen protons in the nucleus. This is what makes it potassium. But how many neutrons are there in the nucleus of potassium? The most common isotope of potassium is potassium-39 (you will see this written as ^{39}K where 39 tells you how many protons plus neutrons are in the nucleus and K means that it is potassium with nineteen protons). So, ^{39}K has nineteen protons and twenty neutrons. This potassium is essentially stable and makes up about 93 percent of all the potassium in a banana. Most of the rest of the potassium in the banana is ^{41}K, and it is also stable. That leaves just a tiny fraction (0.012 percent) that would be ^{40}K, and this potassium is radioactive.

The ^{40}K can decay through three different paths. The most common decay is beta decay. In this case, the potassium produces an electron and the nucleus becomes a calcium atom. The next most common path is through electron capture. A free electron interacts with the nucleus and it becomes argon. Finally, a small fraction of these are beta decays that produce a positron, and the nucleus again turns into argon.

Let's be clear: A positron is the anti-matter version of an electron. Yes, anti-matter like in *Star Trek*. That means it has the same mass as a normal-matter electron but an opposite electrical charge.

If you crunch the numbers (and someone already has[27]), a normal-sized banana on average produces one positron every 75 minutes.

If I were to summarize the banana so far, it would be: mutant, clone, radioactive, and anti-matter.

Now for something different. Could someone use bananas to make a nuclear-powered generator? If so, how many bananas would this generator need to run my house? First, how does a nuclear-powered generator work? Basically, it just uses the energy from the nuclear reaction to boil water. The steam from this boiling water is then used to turn a turbine connected to an electric generator. This is exactly the same way a coal power plant works except that it uses burning coal to heat the water.

Here is my version of a generator: I will start with a spherical mass of bananas. Surrounding this I will have a thin shell of water. As the bananas produce positrons, these positrons will annihilate electrons and produce energy. This energy will then boil the water and create steam to turn the turbine. It's so simple that I don't understand why these things don't already exist. Yes, I know it is actually more complicated than this. I am assuming that my water will be thick enough to not let the positrons escape. Also, I am ignoring the energy from the other types of radioactive decay. Why would I ignore the other decays? Because saying it is a banana-powered anti-matter generator is just cool. Saying it's a generator running off the food decay of bananas is just gross.

The real question remains: How many bananas will I need? Let's look at the power of just one banana. If it produces one positron every 75 minutes, I could calculate the power as the energy from this positron divided by a time of 75 minutes. When a positron annihilates, the energy comes

---

27 http://tertiarysource.net/ts.cgi/anti-banana

from the mass of the electron plus the mass of the positron. The amount of energy would then be mc^2 where c is the speed of light. For one annihilation, we get 1.64 x 10^{-13} joules. That's not much energy. If this much energy is produced every 75 minutes, we would have an average power of just 9.11 x 10^{-18} watts.

Now all I need is to determine how many of these tiny watt bananas I would need to get 2,000 watts (which is about how much I would need to run anything in my house). 2,000 watts divided by the power of one banana tells us that we would need 2.2 x 10^{20} bananas to run the house.

Ok, that's a lot of bananas. How big would my banana generator actually be? Let me just make some estimates. Say that an average banana is 150 grams (0.15 kg) with a density of about 1 g/cm^3 (1,000 kg/m^3). The mass of all these bananas would be (0.15 kg)(n) = 3.3 x 10^{19} kg. With the density, I can calculate the volume of all these bananas to be 3.3 x 10^{16} cubic meters. Smashed into a ball this "banana ball" would have a radius of 2 x 10^5 meters and could easily be visible from space.

Maybe this wasn't such a good idea. I doubt there are enough bananas on the Earth to build my generator. Maybe I would be better off using the bananas to feed some monkeys and have a monkey-powered generator.

I WOULD RATHER FIGHT A HORSE-SIZED DUCK

This is a famous question that is often asked online:

> *"Would you rather fight a horse-sized duck or 100 duck-sized horses?"*

I asked my kids and they had a very interesting discussion. For me, the answer is the horse-sized duck. Now I will tell why physics helps me make this decision.

First, how big is a horse-sized duck? That right there is a tough question. What kind of horse is it? What kind of duck? Are the duck and the horse the same height or the same mass? Is it the same height as a duck in the normal, upright position or with the neck stretched out? No one knows.

When I think of a duck, I always think of a common mallard duck. According to Wikipedia, the mallard is 50–65 cm long with a mass of 0.72–1.58 kg. However, I'm not sure what "long" means. Is that the duck stretched out or in a normal position? Wikipedia also lists the bill as having a length of about 4.5 cm. Using a picture of a duck, I get a height of about 27 cm.

Well, what about a horse? Yes, Wikipedia has a page on horses too. Referencing that, I am going to say an average horse has a height of 1.57 meters at the shoulders and a mass of 500 kilograms.

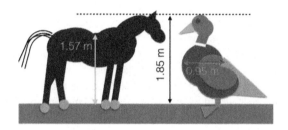

What if the duck and the horse are the same "size"? I am going to say that the duck and the horse would have the highest point of their heads at the same vertical level as shown in the image above. From this image I also estimated the width of the huge duck.

Why do I need that horizontal measurement for a horse-sized duck? I need to somehow get an estimate of the duck's mass. Let's assume our normal duck is like a sphere, a spherical duck. In this case, I would estimate it to have a spherical radius of about 7 cm. With the mass of 1.2 kg (I picked a value from the suggested range) I can calculate the duck density as 835 kilograms per cubic meter.

Two points: that isn't the real density of the duck. Remember, I used the whole mass of the duck but just the spherical body for the volume. Second, even with this method the density is less than water (1,000 kg/m³). This means the duck would float. Everyone knows ducks, wood, gravy, and very small rocks float.

Now, if I use this same duck density I can calculate the mass of the horse-sized duck. Should a horse-sized duck have the same density? I don't know. I didn't come up with the question. However, if you want your horse-duck to float it would need a density less than water. Using the volume of a horse-sized sphere and the density of a duck, I get a mass of 3,000 kilograms.

That is a massive duck, right? Why is it six times more massive than the horse? If you look at both a duck and a horse from the front you will see why. When compared to the duck, the horse is much thinner in comparison to its height.

From this you could say, "Don't mess with that massive duck!" But wait, I said I would fight the horse-sized duck, right? Why did I say that? I don't think this duck could move. Well, it's clear that a duck this big couldn't fly, but I don't think it could even move.

It doesn't matter if the duck can fly or not. What matters are the duck's legs. A duck-sized duck has two approximately cylindrical legs. Looking at the duck image, I measure a leg radius of about 0.005 meters. What is the compression pressure in these legs for a normal duck? It would be the weight of the duck divided by the total cross-sectional area. Let me just use one leg because at some point (like for a walking duck), the duck would have to put all the weight on just one leg. This pressure would be 150,000 newtons per square meter.

That might seem like a lot of pressure, but it isn't. Bones can have a compressive strength about 1,000 times higher than this. So, ducks can walk. However, we already knew this.

If we ramp this up to our horse-sized duck, what happens? The mass increases and so does the radius of the leg. The horse-sized duck is 6.85 times larger than a duck. The leg would also be 6.85 times larger. This would give a horse-duck compression pressure of 8 million newtons per square meter. That is close to a hundred times the pressure of a normal duck. I'm not saying this would definitely break the bone, but come on. This should make walking pretty tough. I think this duck would just sit there quacking, but in really loud quacks. I could just toss some rocks at it until I was declared the winner.

But what if I interpreted the question incorrectly? What if the duck is not the same height as a horse but the same mass?

Let's say the duck is 500 kg like a horse. If it has the same density as a normal duck and I pretend it is a sphere, then the radius of this sphere would be 0.52 meters. This is close to the size of an ostrich. In this case, I would rather fight a hundred duck-sized horses. From what I understand, it isn't a good idea to mess with big, flightless birds. I'm not sure about Big Bird (from Sesame Street), but I always assumed he was a pacifist.

CHAPTER 9: SCIENCE FICTION

HOW MANY ZOMBIES COULD YOU DRIVE THROUGH?

Great questions come from great people. Unfortunately, in this case I don't recall where this question came from. If I had to guess, it was someone in the Global Physics Department meeting. Since I am possibly taking the question from that meeting, let me give them a shout-out. You can find out more about the Global Physics Department at http://globalphysicsdept.org.

Now, back to the question. It seems that everyone loves questions about zombies. I don't know why, but there you are. You are the lone survivor with hordes of zombies around you. There is a car with gas but zombies fill the road. It's like a zombie street party. Can you make it through them? Only physics can save you.

How can you approach a problem like this? It seems almost too crazy to start. Think about what is happening when a car hits just one zombie on the road. The car will push on the zombie and increase its speed. Since forces are inter-actions between two objects, this means the zombie pushes back on the car. During this collision, the forces between the zombie and the car have the same magnitude but different directions. Also, the zombie pushes on the car for the same time interval that the car pushes on the zombie. How could these times even be different? They couldn't.

Since you know the force and time on both the car and the zombie are the same, they have to have the same change in momentum. In the case in which the zombie sticks to the car, the momentum of the car plus zombie after the collision would have to be the same as the momentum before the collision since the zombie wasn't moving. You could think of the mass of the car increas-ing since there is now a zombie stuck on it. The only way to keep the momen-tum of this car and zombie the same as before is to decrease the velocity.

So, hitting a zombie with a car would cause a decrease in the speed of the car. The only problem is that this is wrong. There is one other thing to consider, the fact that the car is not just rolling but instead has an engine. This means there can be another force pushing the car forward if the driver is giving the engine gas.

Let's talk about cars without zombies for a moment. First, how does the engine push the car forward? Well, it technically doesn't. The engine pushes on the wheels and the wheels interact with the road. You could say it is the road that pushes the car forward. Just imagine the case where a car is on an icy road. In this case, you have the same engine but the car stays in the same position. Perhaps it would be best to say that it is the friction force between the tires and the road that pushes the car. If you tried to turn the wheels too fast—so fast that the force needed to accelerate the car was greater than the frictional force—the wheels would just spin.

Why does friction need to push the car anyway? If you have a constant force on an object, the object will keep changing its speed. So, wouldn't you be able to just get up to the speed you want and then turn off the engine? No, that wouldn't work. This would only work in the case where there were no other forces pushing on the car. For a real car, there are two significant forces you have to deal with. First, there is the rolling friction. This is a force that pushes in the opposite direction as the motion of the car, and it is mostly due to tire compression and friction in the axles. The other force is the air drag. The faster the car goes, the greater this force pushes on the car in the opposite direction. With both of these forces, the friction from the tires must continue to push forward for the car to travel at a constant speed so that the net force is zero.

Of course, you already knew this. You know that if you take your foot off of the gas pedal, the car slows down.

Okay, back to zombies. I have touched on two important factors for driving through a crowd of zombies: first, hitting a zombie will slow the car down. Second, the car won't slow down if there is friction pushing it forward. This means if you hit just one zombie, the car might slow down a little bit but it can speed right back up. Unless the zombie does some type of critical damage to the car, everything will be fine.

One zombie collision may be simple, but how would you determine the effect of several zombies? Instead of looking at each individual zombie-car collision, I will adapt another model to this case. Colliding with zombies is very similar to a car colliding with air. The only difference is that the zombies have more mass and are spaced out more.

If you recall, a great model for the air resistance force says that the force is proportional to the product of the density of air, the cross-sectional area of the object (car), a drag coefficient that depends on the shape of the car, and the square of the car's velocity. How would I change this to work with zombies?

The first change is the density. The density of air is about 1.2 kilograms per cubic meter. What about zombies? Let's make some assumptions. A real human has a mass somewhere between 50 and 70 kilograms. What about zombies? I would guess that a zombie's mass would be a little bit smaller since they tend to lose body fluids and appendages rather easily. I am going to guess an average zombie mass of 60 kilograms. If there are "n" zombies per square meter of road, this would give an area density of (60 kg)*n zombies per meter squared. Oh, you don't like that? You are upset that the density is in kilograms per square meter instead of kilograms per cubic meter? I understand your frustration, but it will work out. Trust me.

Here is where I fix the density problem. I know that the zombie drag force should have units of newtons, just like any force. If I use a zombie density in units of kilograms per square meter, the air resistance equation won't work. It won't have the right units. However, what if I change the cross-sectional area of the car to the linear width of the car? If you are hitting a zombie, you won't hit any more if your car is taller. All that really matters is the width. This change will also fix the unit problem. The density will be in kilograms per square meter and the width will be in meters. Our new zombie-drag expression will have the same units as the air-drag expression. Both will be in newtons.

There is still one more thing to take into account: the drag coefficient. For objects moving through air, this coefficient takes into account the aerodynamic shape of the object. A cone-shaped object will have a lower drag coefficient than a flat circle even though both shapes look like circles when looking straight at them. How do you come up with a drag coefficient? In most cases, this is done experimentally. Fortunately for us, we don't have hordes of zombies to run crash test experiments on. That's okay, we can just guess at a coefficient. If we had a coefficient of zero, there would be no interactions with the zombies. A coefficient of one would be similar to inelastic zombie collisions with all the zombies stuck to the car. I always pictured zombies as sort of sticky, but not completely. Let's go with a zombie-drag coefficient of 0.8.

I guess I need to start making some estimations. I will start with the number of zombies per square meter. One zombie per square meter would be a loosely-packed zombie crowd and I think four per square meter would be the maximum you could realistically fit together.

The only other thing I need to estimate is the width of the car. Let me pick a real car. When I imagine myself in a zombie apocalypse, I see myself in a Toyota FJ Cruiser. I don't know why, it just seems like a sensible car for that situation. If you look this car up online, you can find that it has a width of 1.9 meters. Oh, and the mass is about 2,000 kg. I will need the mass later.

Then what will the zombie-drag force on a car like this be? Well, I need to know a speed. Let me calculate the drag at 25 mph and 50 mph. You wouldn't want to go much slower than 25 mph during a zombie interaction and over 50 mph is just out-of-control fast. For a zombie density of one zombie per square meter at 25 mph, I get a drag force of 5,700 newtons (1,280 pounds). For a vehicle speed of 50 mph, I get a drag force of 23,000 newtons (5,170 pounds). Notice that since all I did was double the speed, the zombie-drag increased by a factor of four because the drag force depends on the square of the speed.

I still haven't answered the zombie question. How many zombies could I drive through? Well, let me first calculate the zombie crowd density that I could drive through at a steady speed. If the friction force pushing the car forward is equal to the zombie drag force, the car will just keep moving at a constant speed. So, what is the maximum frictional force that I can push the car forward with? The typical model of friction says that the maximum static friction force is equal to some coefficient of friction multiplied by the force that pushes the two surfaces together. We commonly call this the normal force because it is perpendicular to the surface. If the car is on a flat road, the normal force would just be the weight of the car. Also, for typical tires on a typical road, the coefficient of friction would be about 0.7 for dry roads. Using the mass of the FJ Cruiser, I get a maximum frictional force of 13,700 newtons.

If I assume a driving speed of 25 mph, I can solve for the number of zombies per square meter to balance this force. Using the same values above, I get 2.4 zombies per square meter. That's not too bad. I think I could do that. Of course, if I increase the speed to 50 mph, I am going to have a problem. In that case I can only drive through 0.6 zombies per square meter. Yes, that is a lower value for a higher speed. Why? Because this is the calculation for a car going at a constant speed through an infinite number of zombies. If you are driving faster, you are both having more collisions and bigger collisions.

Okay, one more idea. What if there are not an infinite number of zombies? What if it the local high school football team turned into zombies and is blocking the road? How many could you get through? If you assume a zombie density of three zombies per square meter, you can still get through. Here is how: just drive fast. If the group isn't infinitely long, you will start to slow down as you collide with zombies. However, as long as you get to the end without going below some vulnerable speed (maybe 10 mph), you would survive.

Now, the question is: how would you calculate this? That is a little bit tougher to do. What makes it a more difficult calculation? In this case the zombie-drag force would be changing. As your car slows down, the drag force decreases. You would essentially have to do a numerical calculation to determine the distance you could go before slowing. I personally believe in being prepared for any situation in a zombie apocalypse so I will run this calculation.

Using a zombie density of three zombies per square meter (anything lower and I wouldn't have to slow down past 25 mph) and a starting speed of 50 mph, I get a distance of 20.6 meters (67 ft). This is how far I would drive through zombies until I was slowed down to 25 mph.

How many zombies would this be? Let's say they are on a residential-type road with a width of 9 meters. If I drive 20.6 meters, that would put the zombie-occupied area at 185.4 square meters. This, with a zombie density of three zombies per square meter, would give me 556 zombies. That's more than a high school football team. That is like a big high school population. Well, I don't really care who they were. They are zombies now, and I am driving through them. It's all about surviving the zombie apocalypse.

WHY CAN'T WE HAVE REAL HOVERBOARDS?

There are several movies that seemed to have an impact when I was younger. The three *Back to the Future* films are clearly in this group. I will give you a quick synopsis, just in case you have never heard of them. The basic idea is that a scientist creates a time machine. With this machine, Marty McFly goes back in time and accidentally messes up present day events. He then has to go into the future to try to fix things. Okay, that is enough of the plot for now. However, there is just one more important element: in the second movie, Marty and the scientist travel all the way into the future to the distant date of 2015. I think that is funny.

In this future, Marty finds that everyone has hoverboards. These are just like skateboards except they float above the ground. Since it's 2015, we should have hoverboards now, right? Where's my hoverboard?

Recently some people were fooled into thinking hoverboards were real. A video was posted featuring skateboard legend Tony Hawk riding around on a hoverboard. However, if you thought this video was real, here are some tips that should help you question it:

- The video was completely devoid of any technical details of how this hoverboard was supposed to work. It didn't even give a crazy and make believe explanation for how it works.

- The video includes the statement "The following demonstrations are completely real." Is there any better sign of a fake video than outright saying it's not fake?

- The hoverboard in the video looks exactly like the one in *Back to the Future II*.

- The effects in the video make the motion of the hoverboard look just like the hoverboards in *Back to the Future II*.
  In fact, it looks like people are swinging back and forth on a string (because they are swinging back and forth on a string).

- In the video showing several people trying out the "hoverboard," no one fell. Not even once. Everyone just jumped on the hoverboard and started ripping it up.

So you see, there is no need to look at the video in closer detail to show that it's fake. It's just fake.

The real question is, could we actually have hoverboards? How would you make this thing work? If you want a hoverboard, there is one thing it must absolutely do. No matter what, the hoverboard has to have some force pushing up on the person such that the net force in the vertical direction is zero. This means that the hoverboard would have to push up with a force equal to the weight of the person plus the board. Now for some ideas on how this could work.

Make it a mini helicopter. What if there were two small helicopter-like rotors that lifted the person? I think this is the most plausible option for a real-life hoverboard. Sure, it would take a lot of energy for some small rotors to lift a real person, but it's the best bet. I think you could do it if you had an awesome power source. But it wouldn't look just like the one in the movie since it would have some fans on it.

Use anti-gravity. What if there is some type of anti-gravity disk on the bottom of the board? Would that work? Well, it would work if there was such a thing as anti-gravity. I'm sure someone has some physical explanation that allows for objects with mass to be repelled from each other rather than attracted to each other. However, the real problem would be a hoverboard that repels the ground but not the person. Maybe this is why the person has to be strapped in.

Use magnets. Unlike gravity, which always attracts, the magnetic force can be either repulsive or attractive. The north pole of a magnet attracts the south pole of another magnet but repels the north pole of another magnet. We all know this, right? Yes, but there are two problems with magnets. First, if you just use normal (even very strong) magnets, they don't levitate in a stable way.

Two repelling magnets tend to flip over and then attract. One way to fix this would be to use superconducting magnets.

The second problem is with the other magnet. Even if you used superconductors, you would need something in the ground over which the hoverboard hovers. That makes this more like a train and less like a skateboard.

Electrostatic repulsion. This has the same problem as the magnets. Similar electric charges repel, but you would need excess charges in both the board and the ground. Another problem is with the air. If you have charges that create an electric field over 3×10^6 volts per meter, the air essentially becomes a conductor. This is what happens when you have a spark of electricity.

Ion Thruster. An ion thruster is a real thing. It uses a large electric potential difference to accelerate positive ions. As these ions are accelerated, they exert a force on a craft to act as thrust. Really, this is the same way a chemical rocket works but it uses fewer particles and lasts longer.

The problem with the ion thruster is the magnitude of the force. Even the best ones just have thrusts in the range of 1 newton (this is much less than the 650 newton weight of a person). The large potential difference is also a problem. It would be difficult to fit one of these in a hoverboard. However, I think this is at least possible to use for future hover technology. We would just need to get them to be more powerful and smaller. That shouldn't be so hard. Look at how big and slow computers were forty years ago. Who would have thought that we would have something like a smartphone in the future?

HOW DO YOU POWER A TIME-TRAVELING DELOREAN?

As long as we are talking about *Back to the Future*, maybe we could look at the DeLorean time machine. One of the famous lines from the movie comes when Marty shows the 1950s Doc Brown a video of the 1980s Doc Brown demonstrating the machine. The video states the time machine needs 1.21×10^9 watts.

> *Doc: 1.21 gigawatts? 1.21 gigawatts!? Great Scott!*
> *Marty: What? What the hell is a gigawatt?*

The best part is the way that Doc pronounces gigawatt. The common method is to use a hard "g" but he says it as "jiga"-watt.

So what is a gigawatt? A watt is a unit of power. What is power? Power can be one of several things. The most often used way to describe it is as the change in energy in a certain amount of time. If energy is measured in units of joules and the time interval is in seconds, the power would be in watts. So, one watt equals one joule per second. Horsepower is another unit for energy where 1 hp equals 746 watts.

What about the "giga" part of gigawatt? Giga is a prefix for units that typically means 10^9. This means that 1.21 gigawatts would be 1.21×10^9 watts. Is that a large amount of power? Yes. Just for comparison, the nuclear power reactors in a *Nimitz*-class aircraft carrier produce 194 megawatts (1.94×10^8 watts).

What does Doc Brown even mean when he is talking about the power requirements for time travel? Well, Doc said 1.21 gigawatts. To me, this would be like asking how much power it takes to make toast. Yes, you could use a 500 watt toaster. However, you could also use a 250 watt toaster but it would take longer. Maybe there is something special about time travel such that there is both an energy requirement and it has to take place over some time interval. That's what I am going to assume.

If I want to calculate the energy required for time travel, I will need the power (which is given) as well as the time. This means I will need to calculate the time it takes to travel in time. Yes, that seems silly. What I actually mean is to consider how long the DeLorean is in the process of using energy in the time-travel process.

After studying a few video clips from *Back to the Future*, there are actually two different time intervals. In the first case, Marty drives the DeLorean up to 88 mph and boom, he goes back in time. If you measure from the frame in which this process starts up until it ends, you get a time interval of 4.3 seconds. But wait! What about when Marty is trying to get back to the future? In this case, they use a lightning bolt to power the car. Based on my estimate from the clip, the lightning only interacts with the car for 0.46 seconds. I guess I will have to calculate the energy required for time travel for both of these estimates.

With the power and the time, it's pretty straightforward to calculate the energy. This would just be the product of the power and the time interval. Using the two time intervals, I get an energy value of either 5.2 billion joules or 556 million joules. That's not too bad.

But how do you get five billion joules? Doc Brown's first choice was to use plutonium. Although he didn't give too many details, I guess he was using Plutonium-239. Plutonium-239 is radioactive, but I don't think that's how it provided energy in this case. Instead, I guess there was some type of fission process that broke the nucleus into smaller pieces. Since the pieces have less mass than the original, you also get energy (E = mc²). I will skip the details, but let's just say that one plutonium atom produces 200 MeV (mega electron volts) in the fission process. This is equivalent to 3.2×10^{-11} joules.

In a typical nuclear reactor (which probably wouldn't use Plutonium-239), this energy is used to increase the temperature of water to make steam. The steam then turns an electric turbine to produce electricity. Clearly, that's not happening here. I'm not sure what's going on, but surely it's not a perfectly efficient process. I am going to say it's 50 percent efficient.

In order to get 500 million joules (the lower limit), I would need 3.1×10^{19} atoms. Because a single Plutonium-239 atom has a mass of 3.29×10^{-25} kg, this would require a fuel mass of just 1.2×10^{-5} kg. That seems possible.

What about a lightning bolt? Could you get this much energy from lightning? According to Wikipedia, a single bolt of lightning can have about 5×10^9 joules of energy. That would be just perfect for the time-traveling machine.

What if you find lightning or plutonium just plain boring? Maybe some AA batteries are more your style. On average, a single AA battery has about 10,000 joules of stored energy. In order to get 500 million joules, you would need 50,000 AA batteries. Of course, this assumes you could completely drain these batteries in a very short amount of time. That would mean they need to have high current output and get super-hot. The plutonium probably works better.

GOLLUM PHYSICS

Is Gollum the coolest character in *The Hobbit*? Maybe, just maybe. Okay, before I go on I should perhaps give a spoiler alert. There, you have been warned. Perhaps I don't need to give a spoiler alert for this story since the book was published over seventy years ago. Would I have to give a spoiler alert before saying that both Romeo and Juliet die at the end of the play? Oops. Maybe I just spoiled that one too.

Okay, now for that spoiler. In the book version of *The Hobbit*, Bilbo has just been separated from the rest of his party inside some tunnels under a mountain.

> *"When Bilbo opened his eyes, he wondered if he had; for it was just as dark as with them shut. No one was anywhere near him. Just imagine his fright! He could hear nothing, see nothing, and he could feel nothing except the stone of the floor."*[28]

From this passage you could recreate this scene in a movie. It would just be a plain, black screen. You would see nothing. In physics terms we call this "super-dark." Soon Bilbo discovers that his sword glows and he uses it as a way to look for his pipe. Hobbits clearly keep their priorities straight in a time of emergency. Go for the pipe first.

How do we see anyway? Suppose there is a rock on the ground you are looking at outside during the daytime. In order to see the rock, light has to reflect off the rock and enter into Bilbo's eyes. Where does this light come from? It comes from the sword. That's from the sword, to the rock, and then to the eye. After that, Bilbo's eyes process the light and send information to his brain where he forms the image of a rock.

These are the two ways that humans see things: they see something because that something gives off light or they see something because that something reflects light. Either way, light has to enter the eye in order to see something. If there is no sword, if there is no light, then there is no seeing. The color black is what our brains give us when we don't see any light.

Let's try an activity. Go find some friends (or make some friends) and ask the following:

"Suppose you went into a room with no windows and the door had a light-tight seal. In the room, there is a red apple sitting on a table. Now someone turns off the light. There are absolutely no other lights other than the one that was turned off. When you look at the apple with the lights off, what do you see?"

These are the responses you will probably get:

- You will just see black, and you won't be able to see the apple at all.

- At first it will be dark, but after some time your eyes will start to adjust and you will see the shape of the apple. It won't be red, though, it will be kind of gray.

---

28 J.R.R. Tolkien, *The Hobbit* (New York, NY: Houghton Mifflin Harcourt, 1978), 83.

Just about all answers will be similar to one of these two responses. I find about 20 percent give the first response and 80 percent say you will see something after your eyes adjust. If you ask the 20 percent how they know that is the answer, most will say they have been in a cave with the lights out. If you have ever done this, it is dark, crazy-dark (which is the same thing as super-dark). Some of this 20 percent have been in another location with no lights, like a dark room for developing film. Good luck finding one of those now.

Why do so many people get this question wrong? Well, it turns out that there is usually *some* light. With even just a little bit of light, you can see something. Go out in the woods at night and you can probably see a little bit. If there is a full moon, you can see quite well. Turn off the lights in your bedroom and you will be able to see. There is probably light from the street that can get through your window shades.

Ok, so that is how humans see. Now let's look at some ways that we could have Gollum see in the dark.

What about owls? How do they see so well at night? Well, some animals have eyes like telescopes. Telescopes do more than just magnify the image of distant objects. They also increase the light-gathering ability of a person. This is essentially what owls do. If you have a bigger eye (or a bigger pupil), you can get more light into your eye so that your brain can process it into an image.

Try this. Find a pair of binoculars and look at the size of the lenses. They are quite a bit larger than your eye, right? Now look at the stars in the sky with your eye and then through the binoculars. You can see many more stars with the binoculars. This isn't because of magnification. Rather, it is because your eyes get more light from the sky.

Would big eyes or something like a pair of binoculars work in a cave? No. This method of "night vision" works by getting more light. If there is no light (and trust me, there is no light in a cave) then there is nothing to gather.

If owl-like eyes wouldn't work, then how about some of those military night vision goggles? I hate to spoil this for you, but these night vision goggles do the same thing as the binoculars (essentially). The goggles have image sensors along with tiny video monitors. The light hits the sensors (just like a video camera), and then the image is processed to enhance it to the point that you can see it. Along with the normal visible light spectrum, these embedded cameras also collect infrared light that is close to the visible spectrum. Actually, most cameras can see things you can't see. We typically call this spectrum the "near" infrared. It's the range that the remote control for your TV uses. Try looking at the remote though different cameras. On some of them you will be able to see the little lights on the front of the remote flashing. Some cameras put an IR filter on the front lens to block this spectrum of light since you can't see it with the human eye.

Would night vision goggles help Bilbo? No, they wouldn't. Night vision goggles use batteries and they don't have those kinds of batteries in Middle-earth.

What if you change the camera so that it sees in the far infrared spectrum? We refer to this as a thermal camera. Why is it called "thermal"? Well, it turns out that everything gives off light. The wavelength of light depends on the temperature of the object. For most objects, this wavelength of light falls in the "far" infrared range. So, a thermal camera detects this range of wavelengths and converts it into a false color image. In the image above, different colors represent objects with different temperatures (well, usually).

The image above shows two kids sitting on the ground. The wall behind them is much darker than they are since it is at about room temperature (maybe 20 °C or 70 °F). However, they are at a much warmer temperature (body temperature), so they look brighter and are a different color. But what about the floor right in front of them? Notice how it is also brighter than the walls? It is probably the same temperature as the walls, but it is reflecting infrared light from the two kids.

Would this work for Bilbo in a cave? Okay, let's assume he found some batteries. It would kind of work. It wouldn't show you any details in the cave walls if all the walls had the same temperature (which they typically do). However, if Bilbo was close, it would detect the reflected infrared light from his body. He could also turn the camera around and use the display as a mini-flashlight. Oh, and he might be able to see warm spots on the ground if a person just walked by. It would depend on the sensitivity of the thermal camera and how recently the footprint makers had passed.

Or he could take the batteries out of the thermal camera and use them with a piece of metal to short the batteries and make a fire.

What about bats? Yes, bats live in caves too. However, they don't usually live far into the cave where it is pitch black. But they do have another method for "seeing" things: echolocation. The basic idea is to use ultrasonic sound. You (as in you, the bat) make these high-frequency squeaks. The sound then goes out and bounces off the environment and comes back to your ears. By listening to this sound, you can figure out how far away objects are based on the time it takes for the sound to come back. I guess you can determine shapes based on the type of reflected sound. Yes, this isn't really "seeing," but it is "sensing."

If Bilbo had thought of this (and practiced beforehand), he could use echolocation. His squeaks probably wouldn't be ultrasonic, however, and the goblins could probably hear anything Bilbo could hear.

So, seeing in the dark is a tough problem to fix. How does Tolkien address this? Clearly, from the above quote from *The Hobbit*, Bilbo can't see in the dark. But what about Gollum? Bilbo wanders around and finds the underground lake where Gollum lives. Gollum watches Bilbo from his island in the lake. How? Well, if Bilbo had his sword out, there could be a little bit of light, perhaps enough for Gollum to see him.

Gollum then decides to have a closer inspection of Bilbo. After a short altercation (where Gollum thinks it might be okay to eat Bilbo), Bilbo puts on the ring and becomes invisible. Gollum thinks that Bilbo is running away toward the exit when, in fact, Bilbo is just sitting there being invisible. After Gollum races past Bilbo, we have this passage:

> *"What could it mean? Gollum could see in the dark. Bilbo could see the light of his eyes palely shining even from behind . . . Still there it was: Gollum with his bright eyes had passed him by, only a yard to one side."*[29]

What does this mean? Who knows for sure? When I read this, it seems to suggest that Gollum can see in the dark because his eyes give off light and Bilbo can also see this light. If Tolkien was younger, I would say he had been influenced by the scenes of darkness in cartoons like *Looney Tunes*. If you have seen Bugs Bunny in the dark, you just see his eyes. But Tolkien came way before *Looney Tunes*. Maybe instead, *Looney Tunes* was inspired by Tolkien.

---

29 J.R.R. Tolkien, *The Hobbit* (New York, NY: Houghton Mifflin Harcourt, 1978), 85.

It's rather interesting that the Gollum method of night vision agrees with common ideas about seeing. People that think you can see something in a completely dark room might have the idea that seeing has to do with your eyes. Well, it clearly has something to do with your eyes. But one common idea is that "something" comes out of your eyes so that you can see. Maybe this is light, maybe it is "vision," or maybe it is something else. But with this idea, eyes are like active sonar or echolocation in that they send out something to actively see.

Yet the idea that Gollum has light coming out of his eyes might seem a little crazy. What other ways could Gollum see in the dark?

First, there is the possibility of infrared eyes. Suppose his eyes can detect both near and far infrared. Like I said above, this would sort of work. He wouldn't be able to see very far, and he would have to use the infrared from his body to see things close by.

What other options are there? What about neutrinos? What if there *is* light in this underwater lake? In this case, light comes from the interaction between neutrinos and the water. What is a neutrino? It is a low-mass particle without any electrical charge. This makes them difficult to detect. One way to detect them is to look for light they create as they pass through materials like water.

The amount of light produced isn't very much, but you can detect it. Maybe in Middle-earth the neutrino flux is so high that there is even more light produced from the water in the lake. Maybe that's why Gollum lives there.

While we are on the topic of Gollum, let's look at another physics situation regarding his life in a cave. Gollum lives on an island in an underground lake, right? He lives on fish and things he can steal from the goblins. How often would he have to eat? That is what I want to look at.

To the novice, this question seems impossible to approach. But if you sneak up on this question carefully, you might be able to get an estimate. This is the difference between an expert at B.S. (me) and normal people. I'm not afraid to try and answer this question.

Where would you start? First, why would Gollum need to eat at all? Well, he needs energy. We mammals (I am assuming Gollum is a mammal) eat, breathe, and drink water (and sometimes beer). Our bodies then take this stuff and use it to give us energy. Gollum needs energy to move around and attack Bilbo, that is for sure. He also needs energy to keep his body temperature up to some reasonable level. We all need to do this. However, Gollum is in an underground environment that is typically at some constant (and cold) temperature. If he doesn't eat, his body would eventually reach the same temperature as the environment around him. I assume this would be bad. For humans, we need to keep our body temperature above 85 °F to function at the non-dying level. The sunlight can also help increase our body temperatures.

To start this problem, I need to get some values. I am going to do this two ways. I will make some guesses, but then I will proceed in a symbolic manner. That way, at the end, I should be able to get an expression for how often he needs to eat fish. This expression can then be recalculated with different starting values. Here are some things we need:

- Cave temperature: T_c. If I had to guess, I would put the temperature of the cave, rocks, and water at 50 °F (10 °C). This is a complete guess. The only thing I have to go on is that Bilbo says the water is cold when he accidentally touches the lake.

- Gollum's operating body temperature: T_g. I have a feeling that hobbits are a lot like humans. Since Gollum used to be hobbit-like, he could have a hobbit-like body temperature. But he isn't a hobbit. He has lived in that mountain for quite a long time. Let's say his body temperature is 85 °F (29 °C)

- Gollum's mass: m_g. Okay, let's assume a hobbit is about 1 meter tall. This is right around half the height of a human. What about the mass though? I would think this would be less than half. Actually, I have done this before with estimating the mass of the Hulk. I will skip the details and just give a value of about 20 kg. I think that is way too low, but I'm sticking with it for now.

- Heat capacity of Gollum: C_g. I guess this would be around that of water with a value of 4.19 J/(g*K).

- Amount of energy from consuming one average-sized fish: E_f. I will have to look this up. This random site (http://www.alfit ness.com.au/) puts the energy from 100 grams of raw fish at 427 Kjoules. That means $E_f = m_f(4.27 \text{ KJ/g})$ where the mass of the fish is in grams. Let's just say the average underground lake fish has a mass of 500 grams. That means that one consumed fish would be 2.13×10^6 joules. See, it isn't so scary to just make crazy guesses. It's fun.

That's enough assumptions for now. I suspect I might need a couple more things, but let's move on. If Gollum didn't eat or keep up the temperature, it would explain the decrease in temperature. So Gollum's temperature must have decreased due to a loss of thermal energy. There are three basic interactions that could account for this energy loss.

First, there is conduction. This is an energy transfer between two objects in contact. Energy is transferred from the object with the higher temperature to the object with the lower temperature. For Gollum, this would mostly be an interaction between him and the air since he is warmer than the air. Since air has a fairly low heat capacity, I am going to pretend this heat conduction with the air is small (or smaller than other types of energy transfer). Oh, but what about the water? Clearly, if Gollum is going to hang around in the water, he is going to lose energy much faster than in the air. I guess this is why he uses a boat to get to his home island. Let's assume that Gollum stays dry.

Next, there is evaporation. If you sweat or have water on your skin, this water can go from a liquid to a gas. This phase change takes energy. And guess where the energy comes from. Yes, it comes from the body and results in a decrease in thermal energy and thus temperature. Ok, I already said Gollum stays dry. If we assume he also doesn't sweat (maybe he uses Secret antiperspirant) then we don't have to worry about evaporative cooling.

Finally, we have radiation. Everything radiates energy. The rate that energy leaves an object depends on the surface area of the object and the temperature. Hotter objects radiate energy at a faster rate. This is the energy transfer I am going to look at with Gollum. Why? Because it's something I can estimate. What about a room with objects all at the same temperature? Do they radiate energy as well? Yes. However, they don't get cooler since they absorb radiated energy at the same time they are losing energy. They are in equilibrium.

What is the rate of energy an object radiates? One model for this energy radiation is from the Stefan-Boltzman Law. This says that the power radiated from an object depends on the product of the surface area of the object and the temperature (in kelvins) to the fourth power. Actually, we can also take into account the energy radiated to Gollum. This will make the power that Gollum radiates proportional to his temperature to the fourth power minus the ambient temperature to the fourth power. The point is that this is something we can calculate.

I have an estimate for the two temperatures. All I need to do is to get an estimate for the surface area of Gollum. Let's say that Gollum is about the shape of a cylinder with a height of 1 meter and a radius of 15 cm. Clearly Gollum isn't actually a cylinder, right? I suspect that if you took all the skin off Gollum and flattened it out, the area would be greater than that of a cylinder. However, a good portion of Gollum's skin-surface area could be radiating energy right back to other parts of the skin. Just think of the arms. If you hold your arms by your side, part of your arm skin is facing the sides of your body. Of course, if you huddle in a ball you can decrease your exposed surface area even more.

I have estimated values for all of these quantities, so let me put them into the Stefan-Boltzman model. With this, I get a radiated power of 117 watts. But what about the fish? The key here is that Gollum wants to maintain a constant body temperature. Let me get this radiated power in a different unit. The common unit for power is joules per second. Let me change this to "consumed fish" per second. If I use the estimate above the joules per fish, this would be 55 micro-fish per second, or 55 millionths of a fish every second.

Does this mean that Gollum is going to sit there and eat a tiny piece of fish every second? No, it doesn't. That just has to be the average rate that he eats fish. I assume Gollum has some fat to store energy in. He at least has a stom-ach, so the digestion of the fish will take some time. How many fish a day is this equivalent to? All I need to do is a unit conversion from seconds to days. I get 4.7 fish per day.

That seems like a lot of fish to find in an underground lake. What are the fish going to eat? Okay, I hear what you are saying. Gollum didn't live on fish alone. He stole food from the goblins (or ate goblins). Let's say that just 25 percent of his food comes from fish. That would still be one fish a day (or about 500 grams of fish a day). Seems high.

What about the fish? Wouldn't a fish have to eat something every day? Well, I am really just making up stuff here. I don't think I have ever seen fish in an underground lake. I have seen a cave crayfish before, way the hell back in a cave. There are two important points here. First, fish and crayfish are not warm-blooded. They run their bodies at the same temperature as the water so they don't need to eat to stay warm. Second, cave crayfish don't eat much. They just find stuff that filters through the ground into the cave. There's not much of this stuff to eat and that's why these cave crayfish are rare. Oh fine, I just made up the part about what these crayfish eat, I really don't know.

CHAPTER 10: COULD THIS BE REAL?

CAN HUMANS FLY LIKE BIRDS, WITH WINGS?

Humans have always wanted to fly like birds. It seems like it could work. Unfortunately, I think we are doomed to remain ground-based animals. Well, that is unless you count powered flight or gliding. We can do both of those.

So why couldn't humans fly? Perhaps the simplest answer is "we are too big." Oh, too big you say? Well, surely we aren't too big. Maybe if we just had bigger wings it would work. Nope, it won't work. Well, probably won't work. This falls into a category of things called "stuff that doesn't scale like your intuition." Or perhaps I should say "bigger things aren't the same as smaller things."

Bigger muscles are stronger, but they also have to support more weight. In a rough muscle model, the strength of a muscle is proportional to the cross-sectional area of the muscle. If you double the size of a muscle (and it stays in the same proportions), the strength would increase by a factor of four. However, the weight would increase by a factor of eight.

Perhaps you can start to see the problem. Bigger things get heavier much faster than they get stronger. This is also why my eight-year-old daughter can do more pull ups than I can. She thinks she's tough, but can she drive a car? No, I drive the car.

Let's get back to flying. I suspect that humans are at that size where it would just be super-awkward to have a body build suitable for flying. Perhaps this is also why you don't see giant birds.

How about I look at some birds that fly? You might also think of giant flying dinosaurs (that aren't actually dinosaurs) like the pterosaurs. The largest of these was perhaps the quetzalcoatlus. It seems the size of this ancient flying animal is not universally agreed on. It perhaps had a wingspan of thirty-five feet and a mass from 70–200 kilograms. And did it even fly? Who knows? Maybe it just glided. Maybe it was just big and didn't fly. Perhaps we will never know for sure until we complete our time machine and go back in time to actually observe one in the wild.

What about a bird we know can fly? A real bird that currently exists is the wandering albatross. This is perhaps one of the largest birds alive today with a wingspan of 3.7 meters and a mass of about 12 kg. So here is the plan: browse through Wikipedia and look at birds. It's just like birdwatching, but not birdwatching. For each bird, I will get an approximate wingspan and mass so that I can look for a relationship between these two variables. Please note that all Wikipedia pages for different birds are not the same. They don't all have wingspan and mass information.

Okay, I am going to show you the data in a minute, just calm down. Let me first talk about the Human Birdwings project. Maybe you heard about this some time ago. I will go ahead and summarize the whole story.

In short, a Dutch man makes a series of videos detailing the construction of a set of mechanical wings. Over this time frame, he includes videos of himself testing and eventually flying with the wings. It turns out the whole thing was fake, but a very well-played fake. There, I said it, fake. Well, he said it was fake too (after swarms of Internet users tried to fact-check his story).

Even though it was fake, I will still include both the wingspan and mass of the Human Birdwings project in my data.

I have the data, but what should I plot? I suspect that the ability to fly depends on the size of the wings and the mass of bird (or person). If I plot just the mass versus the wingspan, or even the mass versus the square of the wingspan (since flight probably depends on the surface area of the wings), you will miss a bunch of the data. How can you represent a hummingbird on the same graph as a human with wings? The range of values is quite large.

Here is the best plan: assume a relationship between mass and wingspan. If there is actually a relationship (but it isn't linear), I can assume there is a relationship between the mass raised to some power and the wingspan to some power. If I take the natural log of both the mass and the wingspan, this power will come out as a constant. Yes, I know that sounds crazy, but just check out this plot:

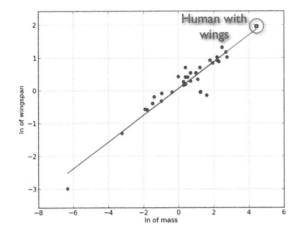

I love it when the data seems to have some relationship.

The cool thing is that the human wings seem to find the rough pattern of wingspan and mass that go along with real birds. Of course, this doesn't mean it would work. This assumes the human has the same muscle distribution as a bird. Typically, we humans don't look the same. If you use your arms for flying, you still carry around these other huge muscles. We like to call them legs. Birds that fly typically have much smaller leg muscles.

In the end, humans can't fly. At least not like a bird. It's not because we don't want to, it's because we are just too darn big.

WHAT IS ARNOLD SCHWARZENEGGER MADE OF?

A long time ago, Arnold Schwarzenegger was a body builder. After that, he was an actor in some pretty iconic movies. One of the movies I remember is *Commando*. Don't worry about the plot. The important thing I want to examine is in just one scene.

I can't even remember why but Arnold, the commando, was chasing some bad guy. He catches the bad guy and then needs to get some important information out of him. What better way to motivate someone to talk than to hold him upside down over a cliff? Yes, he held the bad guy up with just one hand.

Clearly Arnold is strong, but there is more than strength involved here. Here is a sketch of Arnold holding up the guy over the cliff:

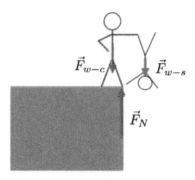

I have included the three important forces on the system consisting of Arnold and the bad guy. Let me assume that Arnold is incredibly strong (which may or may not be true). He is so strong that he can support the victim (his name was Sully in the movie) in this position. This means that I can treat the Arnold-Sully combination as one rigid object. In this case there are essentially three forces on the Arnold-Sully object. There is the gravitational force pulling down on Arnold, the gravitational force pulling down on the Sully part of the object, and the ground pushing up on Arnold. Let me assume he is holding Sully just at the tipping point. If he was any closer to the edge of the cliff, he would fall over.

If the Arnold-Sully object is in equilibrium and not changing in motion, then there are two things that must be true: The net force must be zero and the net torque must be zero.

It may not seem obvious, but most of the time that forces are used, the object is treated like it has no dimensions. Suppose a book is sitting on a table. For this case, we would draw two forces acting on the book: the gravitational force pulling down and the force of the table pushing back up. If I drew these forces, it would be easiest to draw them both on a dot in the center of the book.

Gravity is an interaction between objects with mass. Something like a book is made of many "pieces" of matter, and gravity pulls on all of them. Of course, no one wants to look at all the forces on each little piece of a book. So it turns out we can cheat. All of the tiny gravitational forces on the tiny pieces of the book are equivalent to one single gravitational force pulling on the book at a location we call the "center of mass."

Here is the key: If these pieces are rigid and do not move with respect to one another, both the single force and the four smaller forces do the same thing.

Once I start considering the objects as real objects and not just points, the location that other forces are acting on the object becomes important. For example, imagine I have two Arnold action figures—two objects with the same mass. Both of these objects have the same gravitational force acting on them. However, one has his legs spread wide apart and one is bent at the waist.

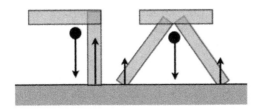

Once I put these down on the table, the total force is zero for both objects. The table pushes up just as much as gravity pulls down. However, the action figure bent at the waist will fall over. Why will it fall over? Once we start talking about rigid objects, we also have to consider how the object rotates. And this is where torque comes into play. If you think of force as the thing that changes the linear motion of an object, torque is like a rotational force that changes the rotational motion of an object. The torque not only depends on what the force applied is, but also where it is applied. Maybe I should start with an example.

Suppose you want to open a door. Typically, to open a door it must rotate. How do you get it to start rotating? Well, one way would be to push on it. Should you push in the middle of the door, near the hinges, or near the handle? If you have opened as many doors as I have, you would know that it is easiest to open a door by pushing on the side away from the hinges (that is why they usually put the handle there). So, by increasing the distance from the point of rotation, you increase the torque.

For the bowing Arnold toy, it should be clear that the torque from gravity does not cancel the torque from the floor. The result is a non-zero torque and the object will start to tip over and fall. For the Arnold in the fighting stance, there are two torques from the floor that do balance the torque from the gravitational force. The net torque on it is zero, so it doesn't change its rotational motion and doesn't fall over.

In general, an object will not tip over if the location of its center of mass is in between its support points. I know I could have just said this without talking about torque, but that wouldn't have been as much fun.

Back to the real Arnold and Sully on the cliff. The ground has to push up with a force equal to the sum of the two weights of the men (Arnold plus Sully). But what about the torque on the system? There are several things that the torque depends on, but for this case we can say there are two things that would create a greater torque: Greater force and greater distance from the pivot point. Think of a seesaw. If Arnold sat on one side of the pivot point and Sully sat on the other, it would rotate Arnold down because he is heavier. If they wanted to seesaw more, Arnold could move closer to the pivot point until the two were balanced. However, Arnold is in no mood to seesaw.

In order for Arnold holding Sully over the cliff to be rotationally stable, the location of the ground pushing up would have to be closer to Arnold than Sully. This is because Arnold has a greater mass, and thus weight. If the torque from the weight of Sully balances the torque from Arnold, he would have to be farther from this point. This way, his smaller weight would be multiplied by a larger distance to give the same torque.

Now for some data. According to Wikipedia[30], Arnold is 1.88 meters tall. By looking at an image from *Commando*, I can estimate that the edge of the cliff to the center of Arnold is about 0.15 meters and the distance from the cliff to Sully is about 0.44 meters.

If I assume Sully is a pretty normal person, he could have a weight around 150 pounds (68 kg). This means in order for the Arnold-Sully system to balance at the edge of the cliff, Arnold would have to have a mass of 199 kg. Oh, you don't like the mass in metric units? Okay, that would be a weight of 440 pounds.

If you compare this weight to Arnold's Wikipedia page, he is listed at 250 pounds. Whoever edited this Wikipedia page used Arnold's volume and the density of a normal person to determine this weight. But who really believes Arnold is made out of normal, human stuff? If I assume that the volume is correct (he does look like a human), this means that in order for Arnold to have this mass he must have a density 1.76 times the density of a human.

A good value for the density of a human is about the same as water (1,000 kg/m³). That would put Arnold's density at 1,750 kg/m³. So, what is he made of then? Aluminum has a density of around 2,700 kg/m³ and titanium has a

---

30 https://en.wikipedia.org/wiki/Arnold_Schwarzenegger

density of about 4,500 kg/m³. Well, I guess he isn't pure aluminum or titanium, though he could be perhaps one-fourth titanium. Maybe he is made of some cool, futuristic material. Remember, he is the Terminator. This just proves it. Now he will never be able to run for President of the United States. I'm pretty sure that to run for president, you have to be a natural-born citizen and human.

The rest of us are not made out of futuristic metal, though the manufacturers of infant car seats like to think we are. I don't mean to attack the infant car seat industry. I am sure some people really like these things. For me, there are only two situations that are good for taking the seat out of the car. First, in a restaurant the seat works quite well with those toddler high chairs. The second useful place for a car seat is in the grocery store. Again the car seat fits right in the standard grocery cart.

In both of the two cases, it is useful to be able to use both hands. For other cases, I find it much easier to just carry the baby. So, what is my problem? What is wrong with these modern devices? It is still about physics and center of mass.

If you have ever carried one of these, you'll recognize the posture in this diagram:

The baby seat has mass and therefore a gravitational force on it. Since the seat cannot be directly over the feet, it will contribute a non-zero torque. This torque is countered by shifting the torso in the other direction. I know you have seen people do this, especially smaller people because they need to move their torso over even further to create an equal and opposite torque. I know what you are thinking, that picture does not look realistic. You are correct. The person should not be smiling. No one smiles while carrying these awkward baby things. Maybe you can see the main problem with carrying this car seat. How are you supposed to walk like that? What if you carried the car seat in front of you with two hands? In this case you would have to walk while leaning backwards. That's still awkward.

Until the seats can get (safely) lighter, or we start making people out of titanium, I recommend purchasing a sling. The advantage of a baby sling is that the mass of the baby is much closer to your feet. This means you don't have to lean the other way so much.

CAN YOU FALL FASTER THAN THE SPEED OF SOUND?

On October 14, 2012, Felix Baumgartner did something incredible. He rode in a helium balloon up to an altitude of 128,000 feet. After that, he jumped and fell for over four minutes before opening his parachute and landing in the open area around Roswell, New Mexico. This was the Red Bull-sponsored Stratos jump. It was a mission similar to a 102,000-foot jump by Joe Kittenger in 1962 and it was awesome.

There are many questions to be answered, so let's get started.

The first question most media outlets asked was: What kind of science do we learn from a jump like this? I don't think this is the best question to ask. It shows a slight misunderstanding of the nature of science. However, I think the late and great physicist Richard Feynman had a good answer to this question (even though he was talking about physics):

> *"Physics is like sex: Sure, it may give some practical results, but that's not why we do it."*

The same is true for the Red Bull Stratos jump. Will we learn something from this jump? Yes, we always learn new things in cases like this. But learning things in this case is just a bonus. We do things like this because we are human. That's just what humans do. We draw pictures, we make music, and we do crazy things like jump out of balloons. We don't need to make excuses for being human, but maybe making excuses for being human is also one of the things humans do.

Now for some physics questions. Did he jump from outer space? I guess the answer to this question depends on how you define "outer space." Perhaps the common definition is "the region where there is no atmosphere." The problem with this is that the Earth's atmosphere doesn't just abruptly end. The transition from space to the Earth's atmosphere is more like a hill rather than a step.

If you want some values, most people consider the International Space Station to be in outer space. This orbits the Earth at about 300 km above the surface (or 980,000 feet). You could say all the way down to 100 km (330,000 feet) is pretty close to space. So, 120,000 feet isn't quite there. Don't get me wrong, it's still way up there.

What is the density of air that high up? On the surface of the Earth, air has a density around 1.2 kg/m³. At 120,000 feet the density is only 7.3 x 10⁻⁴ kg/m³. Because everyone always loves a graph, here is a plot of the density as a function of height:

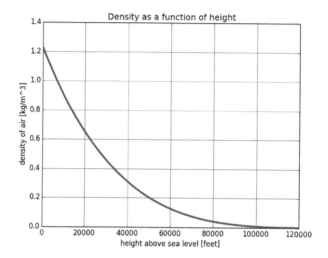

So, if you wanted to call the jumping height "space" it wouldn't be such a terrible thing. You would definitely need a space suit that high, so it could be space, right?

Can you really get to that altitude with a balloon? The first and obvious option for getting that high is a rocket. But why not use some type of airplane? Well, most airplanes need this thing called "air" to work. As you can see from the previous question, there isn't much air up there. So, other than a rocket, the best choice is a balloon. But wait, didn't I just say there isn't much air? Yes, I did. Balloons also need air, but if you get a big enough balloon, you can make it work.

If you think about the balloon at some particular height, there are essentially two forces. There is the gravitational force and a force we like to call the buoyancy force. The buoyancy force is the result of the air colliding more with the bottom of the balloon than it does with the top. The bigger the balloon, the more collisions and the greater the buoyancy force. However, there is a problem. If you just take a balloon and blow it up with air, the gravitational force on the balloon also increases with the size of the balloon. The trick is to use a gas with a lower density than air. In this case, that gas is helium.

Still, as you can see above, the density of air at 120,000 feet is low. With a low density, there are not as many collisions between the air and the balloon. The result is that you need a bigger balloon (which unfortunately has more mass). So, in the end, you need a balloon about 80 meters (over 250 feet) across when fully inflated to lift a jumper and a life support capsule to that height.

Okay, so the air is pretty thin up there, but what about gravity? The gravitational force depends on the distance between the objects (for spherical objects at least). If you double the distance between the center of a planet and a spaceship, the gravitational force will only be one-fourth as much. The key here is "center of the planet." So, if I am 10 feet above the surface of the Earth and I double this height to 20 feet, how far did I move from the center of the Earth? The answer: Not at all (to first approximation). The reason? The Earth is huge. It has a radius of about 6.38×10^6 meters (or 2.09×10^7 feet).

The gravitational force doesn't change too much near the surface of the Earth. But what about at 120,000 feet? Well, a 1 kg mass has a gravitational force of about 9.8 newtons (2.2 pounds) on the surface of the Earth. At an altitude of 120,000 feet the gravitational force would be 9.68 newtons (2.18 pounds). This is 98.8 percent the value at the surface. So the answer is that the gravitational force at 120,000 feet is pretty much the same as on the Earth.

One of the cool things about the Red Bull Stratos jump is that it potentially gave a chance for a human to fall faster than the speed of sound. So, what is the speed of sound? If you think about introductory physics, the speed of sound is often stated as being around 340 m/s or 760 mph. This is the value for the speed of sound at normal temperatures and pressures (like near the surface of the Earth). Unlike the speed of light, it is not a constant. Sound is an interaction between air molecules, so it depends on what they are doing (and it really isn't so simple). However, there is one model for the speed of sound that says it is proportional to the temperature of the surroundings (this is just a model, but it works fairly well).

The higher you go, the lower the temperature (up to a point). Using the same model for the density of air, I can get the temperature and thus the speed of sound. At 120,000 feet, the speed of sound is only around 200 m/s (450 mph).

Can Felix fall faster than the speed of sound? This is the real question you have been waiting for. The answer is yes (probably). To understand how, let's think about the forces on Felix right after he leaves the balloon. Because he isn't really moving yet and there isn't much air anyway, there is just the gravitational force on him. Since this force is acting downward, it causes him to start moving faster and faster as he travels down.

As he starts going faster, there is an air resistance force. You have probably felt this force when you put your hand out the window of a moving car. The faster you go, the faster the force. However, it also depends on the density of air. So, at the beginning of the jump the gravitational force (down) is still greater than the air resistance force (up). This means the net force is still down. Since he is also moving down, this force increases his speed. However, the air resistance force makes the net force a bit smaller so that the rate his speed increases is not as large.

He can't keep speeding up forever. Eventually his speed will get very large and the density of air will increase as he gets lower. At some point, the air resistance force becomes larger than the gravitational force. When this happens, the net force will be upward but he will be moving down. When a net force is in the opposite direction as the motion of the object, the object slows down.

As the fall continues, Felix will slow down to the point where the air resistance force and the gravitational force are the same. When this happens, the net force will be zero and he will neither speed up nor slow down. He will move at a constant speed. This is called terminal velocity.

I am aware that I haven't answered the question. What about the speed of sound? Honestly, finding the speed isn't quite that easy. The best way to do a problem like this is to break it into multiple small steps and let a computer do the work.

Doing that, I made a plot of Felix's speed as a function of time. I have included a curve showing the speed of sound for the altitude he is at during that time:

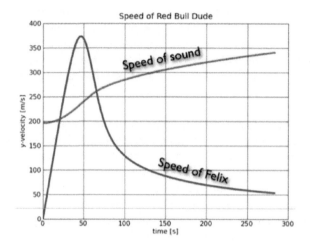

Wait, don't skydivers fall at around 120 mph? Yes, skydivers have a terminal speed around 120 mph. However, Felix isn't the same as a skydiver. As he falls, the density of air (and thus the air resistance) changes. He doesn't stay at the same air density long enough to slow down to a terminal velocity for the first part of the jump. Of course, eventually he will reach terminal speed. However, you can see that according to this calculation, Felix will quickly surpass the speed of sound. He will be a supersonic skydiver.

Ultimately, we don't just have to rely on models and estimations. The Red Bull Stratos jump actually happened and the team measured the speed of Felix during the fall. It turns out that he reached a maximum speed of Mach 1.25: 25 percent faster than the speed of sound.

Well done, Felix. Well done.

CPSIA information can be obtained at www.ICGtesting.com
Printed in the USA
BVOW07*1700060715

407362BV00003BA/3/P